CREWED MOON LANDING FACES MULTIPLE CHALLENGES

U.S. GOVERNMENT ACCOUNTABILITY OFFICE

NIMBLE BOOKS LLC: THE AI LAB FOR BOOK-LOVERS
~ FRED ZIMMERMAN, EDITOR ~

Humans and AI making books richer, more diverse, and more surprising.

Publishing Information

(c) 2023 Nimble Books LLC
ISBN: 978-1-60888-218-2

AI-generated Keyword Phrases

NASA Artemis program; human landing system; space suits; Artemis III mission; progress made by NASA; challenges faced by NASA; lunar landing in 2025 unlikely; ambitious schedule; delays to key events; incomplete orbital flight test; large volume of remaining technical work; contracting approach used by NASA for human landing system and space suits as services.

Publisher's Notes

By 2030, either China or the US, or both, will be on the Moon with permanently crewed bases. This will draw a lot of attention when it happens. *"China lands on the Moon"* would be an epochal headline. Now is the time to start paying attention. This GAO report is a good place to start.

This annotated edition illustrates the capabilities of the AI Lab for Book-Lovers to add context and ease-of-use to manuscripts. It includes five types of abstracts, building from simplest to more complex: TLDR (one word), ELI5, TLDR (vanilla), Scientific Style, and Action Items; four essays to increase viewpoint diversity: Context in the Discourse, Formal Dissent; Red Team Critique; and MAGA Perspective; and Notable Passages and Nutshell Summaries for each page.

ANNOTATIONS

Publishing Information ii
AI-generated Keyword Phrases ii
Publisher's Notes ii
Abstracts iv
    TL;DR (one word) iv
    Explain It To Me Like I'm Five Years Old iv
    TL;DR (vanilla) iv
    Scientific Style iv
    Action Items v
Viewpoints vi
    Grounds for Dissent vi
    Red Team Critique vii
    MAGA Perspective viii
Page-by-Page Summaries xii
Notable Passages xvi

Abstracts

TL;DR (one word)

Artemis.

Explain It To Me Like I'm Five Years Old

The NASA Artemis program is all about sending astronauts back to the moon. They are working on building a special spacecraft called the human landing system that will take the astronauts from their main spaceship to the moon's surface. They are also making special suits called space suits to keep the astronauts safe and comfortable while they explore the moon.

But there are some difficulties they are facing. They have a lot of work to do and some things are taking longer than expected. They also need to do more tests before they do any more.

TL;DR (vanilla)

This document provides an overview of the NASA Artemis program, focusing on the development of the human landing system and space suits for the Artemis III mission. It discusses progress made, challenges faced, and mentions that a lunar landing in 2025 is unlikely due to various reasons. The document also discusses NASA's contracting approach for acquiring these systems as services.

Scientific Style

This document presents a comprehensive overview of the NASA Artemis program, with a specific focus on the development of the human landing system and space suits for the upcoming Artemis III mission. The progress made by NASA and its contractors is discussed, as well as the challenges they currently face. The document acknowledges that achieving a lunar landing in 2025 may be unlikely due to the ambitious schedule, delays in key events, an incomplete orbital flight test, and a significant amount of remaining technical work. Additionally, it highlights the contracting approach used by NASA to acquire services for the human landing system and space suits. Overall, this document provides a detailed and informative summary of the current status and challenges of the Artemis program.

Action Items

Review the progress made by NASA and its contractors in developing the human landing system and space suits for the Artemis III mission.

Assess the challenges faced by NASA in achieving a lunar landing in Evaluate the contracting approach used by NASA to acquire the human landing system and space suits as services.

Identify potential solutions or strategies to address the challenges faced by NASA in the Artemis program.

VIEWPOINTS

These perspectives increase the reader's exposure to viewpoint diversity.

GROUNDS FOR DISSENT

Lack of Safety Assurance: A member of the organization might dissent from this report due to concerns over safety assurance in the Artemis program. They may argue that rushing the development and implementation of the human landing system and space suits increases the risk for potential accidents and compromises astronaut safety. This individual may advocate for a more cautious approach, prioritizing extensive testing and ensuring all technical work is completed before proceeding with any missions.

Unrealistic Schedule: Another reason for dissent could be the belief that the schedule outlined in the document is unrealistic. The member might argue that setting ambitious targets without taking into account potential delays or unforeseen challenges can lead to rushed decision-making, compromising the quality and effectiveness of the mission. They may propose a revised timeline that allows for more flexibility, reducing unnecessary pressure on both NASA and its contractors.

Insufficient Funding: Dissenters may raise concerns about insufficient funding allocated to the Artemis program, which could hinder its success. They might argue that without adequate financial resources, NASA and its contractors will struggle to overcome technical obstacles and meet project milestones. These individuals could advocate for increased funding or a reevaluation of budget allocation to ensure proper execution of the program.

Lack of Transparency in Contracting: Some members might express dissent regarding NASA's contracting approach for acquiring the human landing system and space suits as services. They may contend that relying heavily on contractors while lacking transparency in the selection process could lead to favoritism or subpar performance. These individuals may call for more transparency, accountability, and competition in the contracting procedures to ensure optimal outcomes.

Incomplete Orbital Flight Test: Dissenters might also voice concerns about relying on an incomplete orbital flight test as a basis for assessing progress in the Artemis program. They may argue that without conclusive data from a successful test, it is premature to make definitive statements about the readiness of the program or set firm timelines. These individuals may push for further testing and evaluation before proceeding with any significant milestones or missions.

In summary, a member of the organization responsible for this document might have principled, substantive reasons to dissent from this report due to concerns over safety assurance, unrealistic schedule, insufficient funding, lack of transparency in contracting, and reliance on an incomplete orbital flight test. These individuals may emphasize the need for caution, flexibility, proper funding, transparency, and thorough testing to ensure the success of the Artemis program.

RED TEAM CRITIQUE

The document provides a comprehensive overview of the NASA Artemis program, specifically focusing on the development of the human landing system and space suits for the Artemis III mission. It effectively highlights both the progress made by NASA and its contractors as well as the challenges they face. The information provided is clear and concise, allowing readers to gain a good understanding of the current status and issues surrounding this program.

However, there are several areas where this document could be improved in order to provide a more accurate assessment of the situation. Firstly, it states that a lunar landing in 2025 is unlikely due to an ambitious schedule, delays to key events, an incomplete orbital flight test, and a large volume of remaining technical work. While it is important to acknowledge these challenges, it would have been beneficial if specific details regarding these delays were provided. Additionally, providing insights into how these delays impact overall timelines would have added further clarity.

Furthermore, while it mentions that NASA used a contracting approach to acquire both the human landing system and space suits as services, no explanation or analysis is given regarding why this approach was chosen or what advantages it may offer. Including such information would have deepened readers' understanding of NASA's strategic decision-making process.

Additionally, even though progress made by NASA and its contractors is highlighted in this document; further analysis on whether this progress aligns with initial plans or goals set forth by Artemis program should be included. This will enable readers to evaluate if any deviations from original objectives exist.

Lastly, this document lacks alternative perspectives that could potentially arise from different stakeholders involved in or observing Artemis Program project. Therefore, it does not allow for balanced debate when assessing risks, challenges faced currently, and potential mitigation measures adopted so far. This aspect limits analytical credibility necessary for red team evaluation

Overall, the document provides valuable insight into Artemis program but requires additional information, detailed analysis accounting for challenges faced so far, alternative viewpoints arisingfrom different stakeholders' perspective, and examination of progress made compared to initial goals. Addressing these issues will greatly enhance the comprehensiveness and objectivity of the analysis provided.

MAGA PERSPECTIVE

This document is just another example of the incompetence and failure of NASA under the Biden administration. The fact that they are already acknowledging that a lunar landing in 2025 is unlikely shows just how poorly managed this program is. The American people were promised a return to the moon by 2024, and now it looks like that promise is being broken.

It's clear that NASA is more interested in political posturing and virtue signaling than actually achieving meaningful results. They talk about progress made by their contractors, but where is the evidence? We haven't seen any tangible advancements or breakthroughs from this program.

The delays mentioned in this document are unacceptable. If private companies like SpaceX can successfully launch astronauts into space on a regular basis, then why is NASA struggling so much? It's clear that they are wasting taxpayer dollars and failing to deliver on their promises.

And let's not forget about the contracting approach used by NASA. By outsourcing key components of the Artemis program, they are putting our national security at risk. We cannot rely on private companies to handle such critical missions. This should be a government-led effort with full control and oversight.

In summary, this document only confirms what many of us already suspected - the Artemis program is nothing more than a political stunt designed to appease certain interest groups. It's time for NASA to get its act together or step aside for private companies who can actually get the job done.

GPT-4 ASSESSMENT

Prompt: Compare the Artemis program to what is publicly disclosed about China's moon landing program.

Program Goals:

Artemis Program: The Artemis program's primary goal is to return humans to the Moon, including the first woman and the next man, by the mid-2020s. Its broader objectives include establishing a sustainable human presence on the Moon by the end of the decade to prepare for future missions to Mars.

China's Lunar Program: China's lunar program has been more focused on robotic exploration, with missions like Chang'e-3, Chang'e-4, and Chang'e-5 for lunar landing, rover deployment, and sample return. However, China has also expressed long-term goals for human lunar exploration, aiming to establish a lunar research station and potentially conduct crewed missions, though specific timelines for these crewed missions have not been as clearly defined as Artemis.

Timelines and Progress:

Artemis Program: Artemis missions have specific timelines, with Artemis I being an uncrewed mission around the Moon, Artemis II as the first crewed mission, and Artemis III aiming for a human landing. Delays and technical challenges have been part of the program, but NASA has been transparent about its progress and setbacks.

China's Lunar Program: China has achieved significant milestones in lunar exploration, especially with its Chang'e series of robotic missions.

The timeline for a crewed lunar landing is less clear, but China's space agency has been making steady progress in developing necessary technologies and capabilities.

Technology and Approach:

Artemis Program: Artemis relies on the Space Launch System (SLS), the most powerful rocket ever built, and the Orion spacecraft for crewed missions. It also involves international partners and private companies like SpaceX (for HLS) and Axiom Space (for space suits).

China's Lunar Program: China's program has largely relied on its own technology and resources, with notable successes in robotic missions. The Long March series rockets have been a key part of China's space exploration efforts, and developments in crewed spaceflight capabilities are ongoing.

International Collaboration:

Artemis Program: NASA has emphasized international partnerships in Artemis, with contributions from the European Space Agency (ESA), Canadian Space Agency (CSA), Japan Aerospace Exploration Agency (JAXA), and others. The Artemis Accords also seek to establish a framework for cooperation on the Moon.

China's Lunar Program: China's program has been more nationally focused, though it has engaged in some international cooperation, such as sharing data from its lunar missions with other countries and international scientific communities.

Transparency and Public Disclosure:

Artemis Program: NASA is known for its transparency and regular public updates on its missions, including challenges and setbacks.

China's Lunar Program: While there have been public disclosures about its achievements, China's space program is generally less transparent than NASA's, with fewer details available about future plans or challenges faced.

In summary, while both programs aim for lunar exploration, the Artemis program is currently more focused on crewed missions and establishing a sustainable human presence, with a clear and publicized timeline and international collaboration. In contrast, China's lunar program has seen significant success in robotic missions and is steadily progressing towards its long-term goals for human exploration, but with less specificity regarding timelines and a more nationally focused approach.

Page-by-Page Summaries

BODY-1   *The report discusses the challenges faced by NASA's Artemis program in achieving crewed moon landings.*

BODY-2   *NASA's Artemis III crewed moon landing mission is facing multiple challenges, including an ambitious schedule, delays to key events, and a large volume of remaining work. The mission is unlikely to occur in 2025 as planned.*

BODY-3   *SpaceX has made limited progress in developing the technologies needed for its plan to transfer propellant to a depot in space for the Artemis III mission. Axiom is facing design challenges with their space suit and may need to redesign certain aspects, potentially delaying its delivery. NASA has implemented additional processes and clauses in contracts to ensure mission requirements are met and crew safety is maintained.*

BODY-6   *NASA plans to return astronauts to the moon by 2025 through the Artemis III mission, aiming to establish a sustainable lunar presence and eventually travel to Mars. The original goal of a 2024 landing was delayed due to issues with the human landing system contract.*

BODY-7   *This page discusses NASA's Artemis programs, specifically the development of the human landing system and space suits for the Artemis III mission. It highlights the need for progress in key systems and ensuring contractor compliance with NASA's mission needs and crew safety.*

BODY-8   *NASA's Artemis program aims to return astronauts to the moon and eventually explore Mars. The program involves developing complex systems for missions, including the uncrewed Artemis I and crewed Artemis II missions. The upcoming Artemis III mission focuses on developing a human landing system and space suits for lunar exploration.*

BODY-9   *NASA is developing the Gateway, a lunar orbiting outpost, to serve as a habitat and communication relay for astronauts during Artemis IV and future missions.*

BODY-11   *NASA's Artemis III mission is supported by key program offices including the HLS program, EHP, and the Moon to Mars program office. NASA has awarded contracts to SpaceX and Axiom Space for the development of the human landing system and space suits.*

BODY-12   *NASA is using a service model contracting approach for its Artemis missions, including the acquisition of human landing systems and space suits. This approach increases competition, innovation, flexibility, speed, and affordability. It was previously used for NASA's Commercial Crew Program.*

BODY-13   *NASA evaluated and certified SpaceX for human space flight to and from the ISS, while Boeing is still working towards certification. NASA took steps to reduce risk and increase industry participation in the programs, including conducting risk reduction studies and sharing space suit designs. The human landing system will provide crew access to the lunar surface, with SpaceX developing a commercial Starship vehicle for transportation.*

BODY-14   *SpaceX's HLS Starship system, which includes a booster and crew vehicle, will be used to land NASA astronauts on the moon. The mission involves launching a propellant depot, transferring propellant to the HLS Starship in low-Earth orbit, docking with Orion, descending to the lunar surface for a 6.5-day stay, and returning to Earth.*

BODY-15   *The page shows a figure of SpaceX's mission concept for the NASA Artemis Programs' Human Landing System (HLS).*

BODY-16 NASA has decided to acquire modernized space suits from industry instead of developing them in-house after 14 years and $420 million spent on next-generation space suit development. Axiom is leveraging NASA's government reference design for lunar surface space suits and associated systems. NASA oversees the contractors' development of the human landing system and space suits through a life-cycle review process.

BODY-17 The page discusses NASA's life cycle for space flight projects, including phases such as cost estimation, technology development, and design reviews. The HLS program is approaching a key review while the space suit project is nearing a preliminary design review.

BODY-18 The 2025 crewed lunar landing mission is unlikely to happen due to delays and the amount of remaining work. SpaceX and NASA are aiming for a faster development time, but they are achieving key events at a slower pace compared to other NASA projects. The complexity of human spaceflight makes it unrealistic to expect the program to be completed significantly faster than average.

BODY-19 The page discusses the schedule used by NASA's Artemis Programs to achieve key development milestones, comparing it to the average for a NASA major project.

BODY-20 The NASA Artemis Programs, specifically the Human Landing System (HLS) Program, used a greater schedule percentage to achieve planned key reviews compared to other NASA projects. However, SpaceX has delayed several future program events, compressing the schedule and requiring critical demonstrations and reviews to occur in the next 2 years for the planned Artemis III mission in 2025.

BODY-21 SpaceX and NASA are making progress on the human landing system, but have experienced delays. SpaceX completed 20 milestones early, but overall, eight key events were delayed by 6-13 months. This leaves NASA with a short amount of time to ensure safety requirements are met before the Artemis III mission. SpaceX's Orbital Flight Test was incomplete due to a fire, leading to a mishap investigation by the Federal Aviation Administration.

BODY-22 SpaceX's incomplete Orbital Flight Test led to delays in key test events for NASA's Artemis Programs. The test failure resulted in corrective actions for SpaceX and hindered progress on the human landing system technology maturation plan. Reaching orbit is crucial for Starship development.

BODY-23 NASA's Artemis program faces uncertainties and challenges in meeting its lunar landing goal by December 2025, including the readiness of the HLS Starship and unresolved technical issues with the Raptor engine and on-orbit propellant transfer technology.

BODY-24 SpaceX has made limited progress in developing key systems for the human landing system, including docking sensors and propellant measurement. The timing of further tests is dependent on successful preceding flights. Axiom has made progress in developing space suits but still has significant work remaining.

BODY-25 Axiom is progressing in the development of space suits for NASA's Artemis III mission by leveraging NASA's prior work and making modifications to reduce costs. They are repackaging the life support system and building new components to improve the government reference design.

BODY-26 An illustration of Axiom's space suit and its major system components for NASA's Artemis programs.

BODY-27 Axiom is modifying components of the government reference design for NASA's Artemis Programs to meet requirements and address obsolescence issues. They are

BODY-28   working on emergency life support requirements and parts obsolescence, incorporating new technologies and making improvements to the design.

BODY-28   Axiom's proposed design for NASA's Artemis Programs has unresolved design issues and critical technologies that are not yet mature. The use of different manufacturers and design changes have resulted in lower technology readiness levels, requiring further testing and reassessment.

BODY-29   Axiom's development and procurement of space suit components for the Artemis III mission may face potential delays, risking a compressed delivery window. Axiom is working to mitigate supply chain issues by producing some parts in-house. The necessary testing facilities for certifying the suits may not be available in time, requiring alternative testing plans. NASA is also addressing cross-program risks related to integrating the lander and space suits with systems needed for the mission.

BODY-30   The page discusses the challenges of integrating software and hardware for NASA's Artemis Programs, particularly the HLS program. It highlights the need for adequate testing to avoid critical software defects and mentions lessons learned from Boeing's Starliner flight test failure.

BODY-31   NASA is addressing the risks of software defects and lunar dust contamination in the Artemis III program. They are working with SpaceX and Axiom to mitigate these risks and ensure the safety of the crew and equipment before launch.

BODY-32   NASA plans to assess whether SpaceX's and Axiom's systems meet requirements and are safe for the Artemis III mission. The requirements include system functionality, performance, and interface with other systems. NASA allowed alternative technical standards in certain areas for both contractors.

BODY-33   NASA's Artemis Programs have accepted alternative technical standards without creating additional risk. Contractors SpaceX and Axiom will verify that their systems meet NASA's requirements, and design certification reviews will ensure compliance. NASA is formulating plans to determine agency readiness for launch.

BODY-34   NASA is developing guidance for certification of flight readiness for future Artemis missions, including the Artemis III mission. This will establish a framework for programs to develop their own plans and define reporting structures. NASA plans to obtain formal human rating certification for the Artemis III mission.

BODY-35   NASA officials are applying guidance and best practices from previous missions to develop the certification of flight readiness process for the Artemis III mission. The process will include multiple flight readiness reviews and will begin around 3 months before launch. However, there are challenges in integrating five systems for the mission, which will require determining the timing and spacing of the reviews. NASA plans to release both the general certification of flight readiness plan and the Artemis III mission supplement in 2024.

BODY-36   NASA's contracts with SpaceX and Axiom for the Artemis programs grant visibility into contractor efforts, including data deliverables and safety reports. Timely review of SpaceX data is crucial as development progresses.

BODY-37   NASA has insight and access into SpaceX and Axiom's activities, including design, development, testing, and safety aspects of their projects. This helps NASA verify technical information and ensure successful milestone reviews.

BODY-38   NASA and its contractors are using insight clauses to gain visibility into the progress of critical technologies and design efforts. Insight activities pose a schedule risk, but NASA has established processes to adjust the level of insight based on ongoing assessment of risks.

BODY-39   NASA's Artemis Programs are benefiting from insight provided by SpaceX and Axiom, but there may be schedule risks and administrative bottlenecks. However, NASA maintains a strong safety culture and collaboration with contractors is defined in the contracts.

BODY-40   The page discusses the collaboration and insight clauses in the contracts between NASA and SpaceX/Axiom for the Artemis Programs. Both companies are leveraging NASA support, with SpaceX requesting up to 60 full-time NASA employees and Axiom requesting up to 25. Collaboration areas include micrometeoroid orbital debris, engine development, manufacturing, lighting, and crew training.

BODY-42   The page discusses the objectives and methodology of a review conducted by GAO on NASA's progress in developing systems for landing humans on the moon in 2025. It focuses on the human landing system initial capability and space suits for Artemis III mission.

BODY-43   The page discusses the methodology used to examine the development time frames and risks associated with the NASA Artemis Programs, specifically the HLS program. It includes an analysis of data reliability, scope of remaining work, and steps taken by NASA to ensure mission needs and crew safety.

BODY-44   The page provides information on the objectives, scope, and methodology of a performance audit conducted by GAO on NASA's Artemis Programs. It includes details on interviews conducted with SpaceX, Axiom, and NASA personnel, as well as the review of documentation and plans. The audit was conducted from September 2022 to November 2023 in accordance with government auditing standards.

BODY-47   The Government Accountability Office supports Congress in its responsibilities and aims to improve the performance and accountability of the federal government. It examines public funds, evaluates programs and policies, and provides recommendations to help Congress make informed decisions. GAO's publications can be obtained through their website or by phone.

Notable Passages

BODY-2   "The National Aeronautics and Space Administration (NASA) is preparing to land humans on the moon for the first time since 1972 in a mission known as Artemis III. Since GAO's September 2022 report (GAO-22-105323), NASA and its contractors have made progress, including completing several important milestones, but they still face multiple challenges with development of the human landing system and the space suits. As a result, GAO found that the Artemis III crewed lunar landing is unlikely to occur in 2025."

BODY-3   "NASA plans to take multiple steps to determine whether SpaceX's and Axiom's systems meet its mission needs and are safe for crew. For example, NASA developed a supplemental process—one not required by its policies—to determine whether the contractors' systems meet requirements before the mission. Also, NASA's contracting approach to acquire the human landing system and space suits as services included insight clauses in the SpaceX and Axiom contracts. Program officials stated these clauses ensure that NASA has visibility into broad aspects of the contractors' development work, including anything that could affect the Artemis III mission or crew safety."

BODY-6   "The National Aeronautics and Space Administration (NASA) plans to return U.S. astronauts to the surface of the moon by the end of 2025. This mission, known as Artemis III, will be the first time that crew will land on the moon since the 1972 Apollo 17 mission, and the first time ever that crew will land at the lunar south pole. The Artemis III mission is the third in a series of increasingly complex missions to maintain U.S. leadership in space exploration, build a sustainable lunar presence over the next decade, and ultimately travel to Mars."

BODY-8   "The goal of NASA's Artemis enterprise is to return U.S. astronauts to the surface of the moon, establish a sustained lunar presence, and ultimately achieve human exploration of Mars."

BODY-12   "NASA is acquiring the human landing system and space suits as services, which represents a relatively new contracting approach for NASA. NASA's intent is to transition from the traditional government-owned hardware model to a service model. Using this approach, the programs set high-level requirements and rely on the contractor's innovation to develop and deliver the system. Contractors only receive payment after NASA determines that the contractor has successfully achieved a milestone as defined in the contract. NASA is moving toward using this service model contracting approach for some acquisitions on its Artemis missions because the agency believes that it increases competition, innovation, flexibility, speed, and affordability."

BODY-13   "The human landing system will provide crew access to the lunar surface and demonstrate initial capabilities required for deep space missions. SpaceX is currently developing a commercial Starship vehicle to transport humans and cargo to low-Earth orbit, the moon, and Mars."

BODY-14   "Once the lunar surface activities, including moonwalks, are complete, the HLS Starship will ascend back to near-rectilinear halo orbit, where the crew will transfer back to Orion for their return to Earth."

BODY-16   "Modernized space suits and associated hardware will provide portable life support, as well as tools for crew to use for lunar science and maintenance tasks."

BODY-18 "NASA and SpaceX completed several important milestones for the human landing system since our September 2022 report, but a variety of factors make a lunar landing in 2025 unlikely. NASA officials are currently reviewing the HLS schedule."

BODY-20 "SpaceX has also delayed several future program events that further compress the schedule. Since July 2022, the HLS program office and SpaceX delayed multiple key events from 2023 to 2024, meaning that many critical demonstrations and reviews will need to occur in the next 2 years to support an Artemis III mission as planned in 2025."

BODY-21 "In April 2023, after a 7-month delay, SpaceX achieved liftoff of the combined commercial Starship variant and Super Heavy booster during the Orbital Flight Test. But, according to SpaceX representatives, the flight test was not fully completed due to a fire inside the booster, which ultimately led to a loss of control of the vehicle."

BODY-22 "SpaceX representatives said their Autonomous Flight Safety System initiated the vehicle self-destruct sequence and the vehicle began to break up about 4 minutes into the flight after the vehicle deviated from the expected trajectory, lost altitude, and began to tumble."

BODY-23 "The HLS program will need to complete a significant amount of complex technical work on the engines and propellant transfer technology between 2023 and the end of 2025 to achieve the planned lunar landing goal. In a May 2023 NASA document, NASA officials overseeing the Artemis III mission integration stated that the HLS design maturity, with almost 3 years left before the planned launch, was insufficient. For example, they cited on-orbit propellant transfer as a potential issue because significant technical problems still need to be resolved."

BODY-25 "Axiom is also making progress on suit development by leveraging NASA's prior work. Axiom representatives said they brought in relevant experts and personnel who worked on the government reference design. These representatives said their approach was to adopt the government reference design and refine it to reduce costs to NASA."

BODY-27 "Axiom representatives told us they may redesign applicable portions of the suit because the NASA government reference design did not satisfy the requirement to make the suit capable of storing that amount of oxygen. Axiom staff plan to decrease the size and rearrange components in the life support system package design to accommodate larger tanks that can hold more oxygen."

BODY-28 "In January 2023, NASA assessed two critical space suit systems, the Life Support System and the Pressure Garment System, at a technology readiness level (TRL) 4. Axiom's technology assessment rated over half of its critical technologies below TRL 6 and the lowest among those items at a TRL 3. One of those components, the Regenerable CO2 Scrubber, is a life support system subcomponent that removes carbon dioxide from the suit environment. Axiom rated it at a low TRL since Axiom is not using the government reference design for this component."

BODY-29 "Artemis III mission. To support technology maturation efforts, Axiom personnel are developing multiple test rigs for different components of the life support system."

BODY-30 "Integration of software developed for dissimilar hardware platforms, using different operating systems, as well as use and integration of heritage software, can be challenging and prone to introducing defects. HLS risk documentation states that adequate test facilities and test campaigns are required to avoid late discovery of critical software defects because critical issues are often uncovered when software is integrated and tested with flight hardware. Therefore, without adequate testing, it is

	possible that critical software defects are missed. This situation could result in cost and schedule effects, or worse, potential loss of mission or crew."
BODY-31	"NASA plans to take multiple steps to determine whether SpaceX's and Axiom's systems meet its mission needs and are safe for the crew. The HLS and EHP programs will ultimately determine whether the contractors' systems meet contract requirements. Then, NASA will conduct a to-be-decided series of reviews to determine whether the agency is ready for launch based on guidance it is currently developing."
BODY-32	"NASA officials said that after they completed the adjudication process with SpaceX, approximately 50 percent of SpaceX's technical standards were alternative to NASA's."
BODY-35	"NASA officials said they are applying NASA guidance and best practices from earlier human spaceflight efforts to their Artemis certification of flight readiness plans. These earlier efforts include the Space Shuttle, CCP, and ISS missions."
BODY-36	"The HLS program manager said that while the program has yet to experience delays due to needing time to review SpaceX data, program staff recognize that timely review will be critical as SpaceX's development progresses. The program manager said that the program developed a schedule for reviewing SpaceX data to provide its technical opinion on a timely basis."
BODY-37	"NASA officials said that the purpose of ensuring insight into contractor efforts is to verify technical information, which should help ensure that formal milestone reviews and deliverable submissions are successful."
BODY-39	According to both HLS officials and SpaceX representatives, SpaceX gave NASA insight beyond what the contract requires by sharing information about its work under other programs as well as its Starship development. The program reported that its insight into SpaceX's early Starship development has been beneficial for NASA and SpaceX as they solve problems and reduce risk.
BODY-40	"The Axiom contract states that collaboration is the highest form of insight, which EHP officials explained to mean that knowledge gained through collaboration can be used to understand the system and its risks."

United States Government Accountability Office
Report to Congressional Committees

November 2023

NASA ARTEMIS PROGRAMS

Crewed Moon Landing Faces Multiple Challenges

GAO-24-106256

GAO Highlights

Highlights of GAO-24-106256, a report to congressional committees

November 2023

NASA ARTEMIS PROGRAMS

Crewed Moon Landing Faces Multiple Challenges

Why GAO Did This Study

NASA is returning humans to the moon to maintain U.S. leadership in space exploration and prepare for future missions to Mars. NASA is implementing the Artemis missions to meet these goals. To accomplish the Artemis III mission as planned by December 2025, NASA needs to develop, acquire, and integrate several new systems. These include a system to transport crew to and from the lunar surface, and space suits for lunar surface operations. NASA is using a relatively new approach to acquire the human landing system and space suits that is intended to increase innovation and improve affordability. To develop the lunar lander, NASA awarded a contract option to SpaceX in 2021. To develop Artemis space suits, it awarded a contract to Axiom Space in 2022.

A House report includes a provision for GAO to review NASA's lunar programs. This is GAO's fourth report examining the Artemis enterprise.

This report describes the extent to which NASA has made progress in developing key systems needed to land humans on the moon in 2025, and has processes in place to ensure that those systems will meet NASA's needs and be safe.

GAO assessed NASA data, documentation, and policy; analyzed contract documentation, contractor risk charts, and technology maturation plans; and interviewed NASA officials and industry representatives.

View GAO-24-106256. For more information, contact William Russell at (202) 512-4841 or russellw@gao.gov.

What GAO Found

The National Aeronautics and Space Administration (NASA) is preparing to land humans on the moon for the first time since 1972 in a mission known as Artemis III. Since GAO's September 2022 report (GAO-22-105323), NASA and its contractors have made progress, including completing several important milestones, but they still face multiple challenges with development of the human landing system and the space suits. As a result, GAO found that the Artemis III crewed lunar landing is unlikely to occur in 2025. In July 2023, NASA stated that it is reviewing the Human Landing System schedule.

The current challenges that GAO identified include:

- **An ambitious schedule:** The Human Landing System program is aiming to complete its development—from project start to launch—in 79 months, which is 13 months shorter than the average for NASA major projects. The complexity of human spaceflight suggests that it is unrealistic to expect the program to complete development more than a year faster than the average for NASA major projects, the majority of which are not human spaceflight projects. GAO found that if development took as long as the average for NASA major projects, the Artemis III mission would likely occur in early 2027.
- **Delays to key events:** As of September 2023, the Human Landing System program had delayed eight of 13 key events by at least 6 months. Two of these events have been delayed to 2025—the year the lander is planned to launch. The delays were caused in part by the Orbital Flight Test, which was intended to demonstrate certain features of the launch vehicle and lander configuration in flight. The test was delayed by 7 months to April 2023. It was then terminated early when the vehicle deviated from its expected trajectory and began to tumble. Subsequent tests rely on successful completion of a second Orbital Flight Test.

Notional Depiction of the Human Landing System

Source: SpaceX. | GAO-24-106256

- **A large volume of remaining work:** SpaceX must complete a significant amount of complex technical work to support the Artemis III lunar landing mission, including developing the ability to store and transfer propellant while in orbit. A critical aspect of SpaceX's plan for landing astronauts on the moon for Artemis III is launching multiple tankers that will transfer propellant to a depot in space before transferring that propellant to the human landing system. NASA documentation states that SpaceX has made limited progress maturing the technologies needed to support this aspect of its plan.

- **Design challenges:** Axiom is leveraging many aspects of NASA's prior work to develop modernized space suits, but significant work remains to resolve design challenges. For example, NASA's original design did not provide the minimum amount of emergency life support needed for the Artemis III mission. As a result, Axiom representatives said they may redesign certain aspects of the space suit, which could delay its delivery for the mission.

Illustration of Axiom's Space Suit and Major System Components

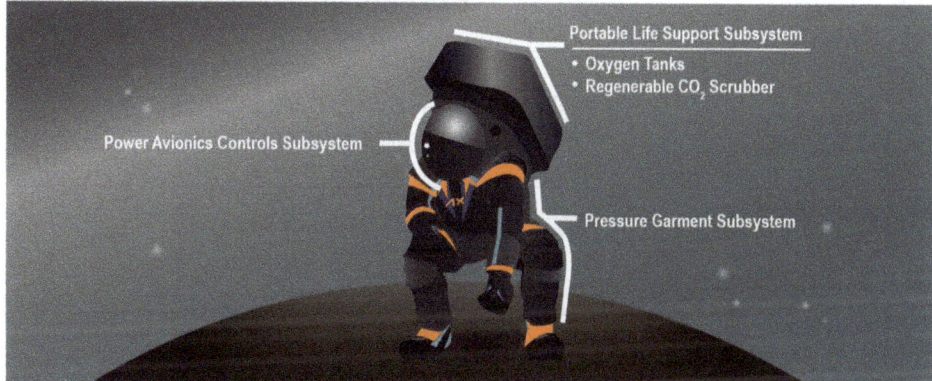

Source: GAO analysis of Axiom information and image. | GAO-24-106256

NASA plans to take multiple steps to determine whether SpaceX's and Axiom's systems meet its mission needs and are safe for crew. For example, NASA developed a supplemental process—one not required by its policies—to determine whether the contractors' systems meet requirements before the mission. Also, NASA's contracting approach to acquire the human landing system and space suits as services included insight clauses in the SpaceX and Axiom contracts. Program officials stated these clauses ensure that NASA has visibility into broad aspects of the contractors' development work, including anything that could affect the Artemis III mission or crew safety. Officials stated that this visibility extends to certain aspects of work SpaceX and Axiom are doing for their commercial endeavors. For example, this included SpaceX's activities leading up to the Orbital Flight Test, which flew a commercial variant of the human landing system.

Contents

Letter		1
	Background	3
	2025 Crewed Lunar Landing Is Jeopardized by Delays and Scale of Remaining Work	13
	NASA Is Taking Steps to Ensure Systems Are Safe and Meet Its Needs before Launch	26
	Agency Comments	35
Appendix I	Objectives, Scope, and Methodology	37
Appendix II	GAO Contact and Staff Acknowledgments	40
Related GAO Products		41

Tables

Table 1: Examples of Requirements Applicable to SpaceX and Axiom, by GAO-Identified Type	27
Table 2: Scope of NASA's Insight into SpaceX and Axiom Activities	32

Figures

Figure 1: Key NASA Programs Supporting Artemis Missions	5
Figure 2: SpaceX Mission Concept for Human Landing System (HLS)	10
Figure 3: NASA's Life Cycle for Space Flight Projects	12
Figure 4: Human Landing System (HLS) Program Used a Greater Schedule Percentage to Achieve Planned Key Reviews than the Average for NASA Major Projects Launched since 2010	15
Figure 5: Completed and Remaining Milestones to Develop Space Suits for the Artemis III Mission	20
Figure 6: Illustration of Axiom's Space Suit and Major System Components	21
Figure 7: NASA Processes to Determine Whether a Contractor's System Meets Requirements before Key Reviews to Support Artemis III Launch	29

Abbreviations

ACD	Artemis Campaign Development
ARGOS	Active Response Gravity Offload System
CCP	Commercial Crew Program
CDR	critical design review
EGS	Exploration Ground Systems
EHP	Extravehicular Activity and Human Surface Mobility Program
EVA	Extravehicular Activity
FAR	Federal Acquisition Regulation
HLS	Human Landing System
ISS	International Space Station
KDP	key decision point
NASA	National Aeronautics and Space Administration
Orion	Orion Multi-Purpose Crew Vehicle
PDR	preliminary design review
SLS	Space Launch System
TRL	technology readiness level

This is a work of the U.S. government and is not subject to copyright protection in the United States. The published product may be reproduced and distributed in its entirety without further permission from GAO. However, because this work may contain copyrighted images or other material, permission from the copyright holder may be necessary if you wish to reproduce this material separately.

November 30, 2023

The Honorable Jeanne Shaheen
Chair
The Honorable Jerry Moran
Ranking Member
Subcommittee on Commerce, Justice, Science, and Related Agencies
Committee on Appropriations
United States Senate

The Honorable Hal Rogers
Chairman
The Honorable Matt Cartwright
Ranking Member
Subcommittee on Commerce, Justice, Science, and Related Agencies
Committee on Appropriations
House of Representatives

The National Aeronautics and Space Administration (NASA) plans to return U.S. astronauts to the surface of the moon by the end of 2025. This mission, known as Artemis III, will be the first time that crew will land on the moon since the 1972 Apollo 17 mission, and the first time ever that crew will land at the lunar south pole. The Artemis III mission is the third in a series of increasingly complex missions to maintain U.S. leadership in space exploration, build a sustainable lunar presence over the next decade, and ultimately travel to Mars.

In March 2019, the White House directed NASA to accelerate its plans for a lunar landing from its original goal of 2028 to 2024, in part to create a sense of urgency in returning American astronauts to the moon. In November 2021, NASA announced that it was no longer working to its goal of an Artemis III lunar landing in 2024 and that the new date would be no earlier than 2025. NASA officials attributed this change to a 7-month delay in working on the human landing system, subsequent to a bid protest and federal court complaint regarding the award of the lander's contract. In announcing the delay, senior NASA officials acknowledged that the prior 2024 goal was unrealistic.

To accomplish the Artemis III mission, NASA is partnering with industry to develop two new systems: the human landing system (HLS)—to transport crew to and from the lunar surface, and modernized space suits for lunar surface operations.[1] In the fiscal year 2024 President's budget request, NASA requested $12.4 billion over the next 5 fiscal years for the human landing system and modernized space suits. In addition to developing these systems, NASA will need to ensure that the human landing system and space suits meet NASA's needs to operate in a deep space environment, conduct scientific exploration, and ensure crew safety.

The House Report 117-395 accompanying the Commerce, Justice, Science, and Related Agencies Appropriations bill, 2023 contains a provision for GAO to continue conducting in-depth reviews of NASA's lunar-focused programs. This is the fourth in a series of GAO reports addressing NASA's Artemis enterprise.[2] The focus of this report is on the human landing system and space suits being developed for the Artemis III mission. This report describes the extent to which NASA (1) has made progress in developing key systems needed to land humans on the moon in 2025, and (2) has processes in place to ensure that its contractors are developing systems that meet NASA mission needs and are safe for crew.

To conduct this work, we reviewed and assessed NASA data, documentation, and policy. For example, we reviewed program plans and quarterly status reviews from the HLS and Extravehicular Activity and Human Surface Mobility (EHP) programs, which are overseeing the development of the human landing system and modernized space suits, respectively. We reviewed these documents to identify program milestones, critical technology demonstrations, and NASA's progress. We also analyzed contract documentation, contractor risk charts, and technology maturation plans to determine the current risks facing the programs. Additionally, we assessed contract and requirements documentation, NASA agendas and presentation slides on lessons learned, and mission integration documentation. We also interviewed a

---

[1]NASA uses the term Exploration Extravehicular Activity system to encompass the garments, interfaces with the human landing system, and tools for the scientific lunar mission. For the purposes of this report, we will refer to the Exploration Extravehicular Activity system as space suits and associated tools.

[2]GAO, *NASA Lunar Programs: Improved Mission Guidance Needed as Artemis Complexity Grows,* GAO-22-105323 (Washington D.C.: Sept. 8, 2022); *NASA Lunar Programs: Significant Work Remains, Underscoring Challenges to Achieving Moon Landing in 2024,* GAO-21-330 (Washington, D.C.: May 26, 2021); GAO-21-105; and *NASA Lunar Programs: Opportunities Exist to Strengthen Analyses and Plans for Moon Landing,* GAO-20-68 (Washington, D.C.: Dec. 19, 2019). For more information, see our related work at the end of this report.

wide range of NASA and industry officials. See appendix I for more information on our objectives, scope, and methodology.

We conducted this performance audit from September 2022 to November 2023 in accordance with generally accepted government auditing standards. Those standards require that we plan and perform the audit to obtain sufficient, appropriate evidence to provide a reasonable basis for our findings and conclusions based on our audit objectives. We believe that the evidence obtained provides a reasonable basis for our findings and conclusions based on our audit objectives.

Background

NASA's Artemis Missions

The goal of NASA's Artemis enterprise is to return U.S. astronauts to the surface of the moon, establish a sustained lunar presence, and ultimately achieve human exploration of Mars. To do so, NASA programs are developing multiple highly complex and interdependent systems that will need to be integrated to support individual Artemis missions.[3]

- The Artemis I and II missions are the first uncrewed and crewed demonstration missions, respectively, of the Space Launch System (SLS) launch vehicle, Orion Multi-Purpose Crew Vehicle (Orion), and associated ground systems, known as Exploration Ground Systems (EGS).[4] Artemis I successfully launched on November 16, 2022, with the Orion capsule safely returning to Earth on December 11, 2022.

- The Artemis III mission incorporates new programs that are developing the human landing system and space suits. The goal of the Artemis III mission is to return U.S. astronauts to the surface of the moon and conduct scientific exploration activities during a 6.5-day stay at the lunar south pole. The characteristics of the lunar south pole affect when and where crew can land. For example, the mission must occur in a location that stays continuously lit, and the crew must have direct-with-Earth communication.

---

[3]NASA distinguishes between programs and projects in its policies and guidance. A NASA program has a dedicated funding profile and defined management structure and may include several projects. Projects are specific investments under a program that have defined requirements, life-cycle costs, schedules, and their own management structures.

[4]SLS is the vehicle NASA will use to launch the Orion Crew Capsule beyond low-Earth orbit. Orion is the crew capsule to transport humans from the Earth to the HLS. EGS is the infrastructure at Kennedy Space Center to support Artemis mission launches.

- For Artemis IV and beyond, NASA is also developing a lunar orbiting outpost known as the Gateway to act as a habitat and safe work environment for astronauts and as a communications relay between the lunar surface and Earth.

See figure 1 for the programs needed to accomplish the Artemis missions.

Figure 1: Key NASA Programs Supporting Artemis Missions

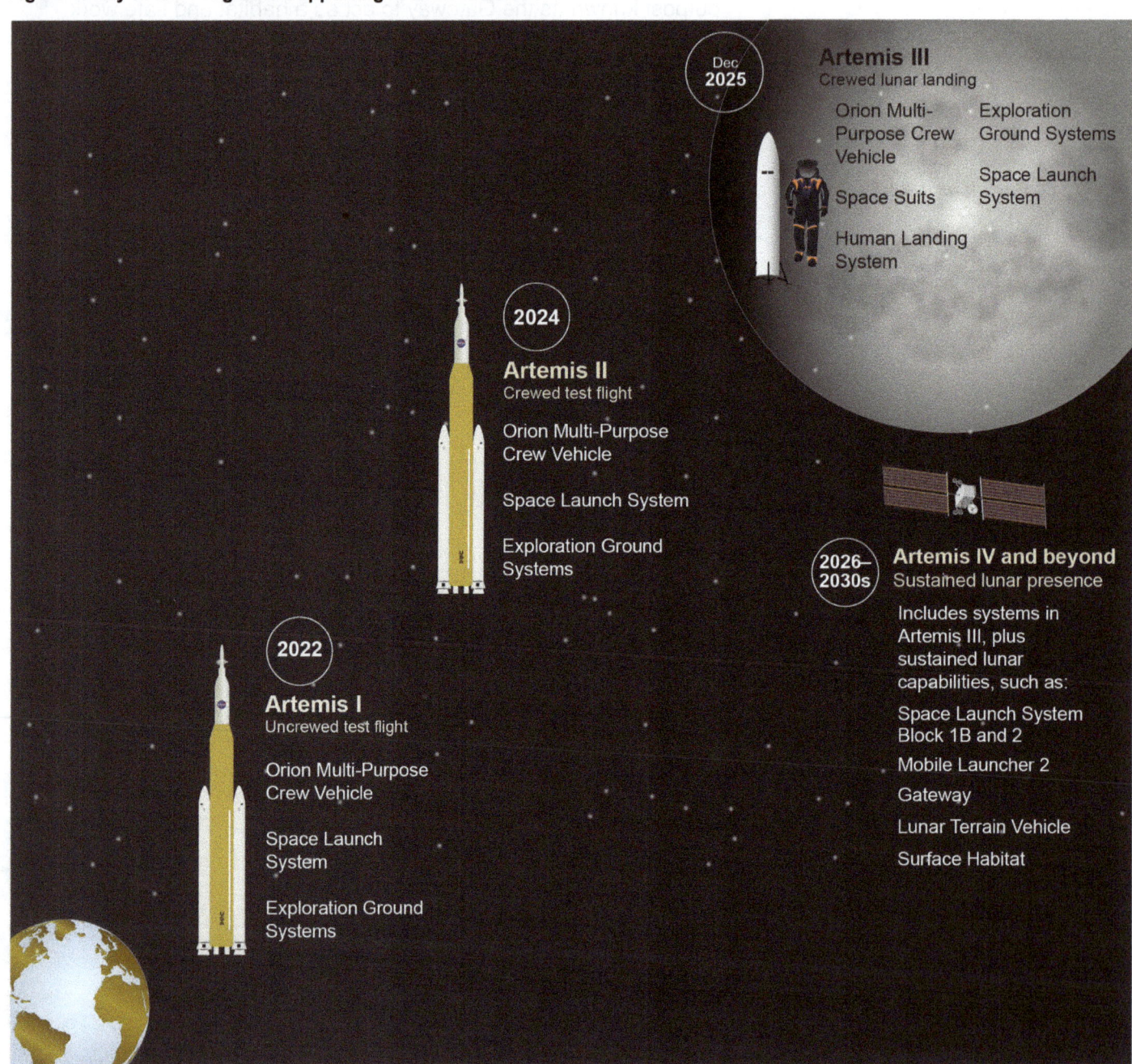

Source: GAO presentation of NASA documentation. | GAO-24-106256

Key NASA Program Offices Supporting the Artemis III Mission	Several key program offices have a role in supporting the Artemis III mission. • The HLS program is responsible for managing the human landing system development and certifying that the integrated lander systems are ready for flight. The program management is located at Marshall Space Flight Center, with key participation from Johnson Space Center and Kennedy Space Center. • EHP is responsible for working with industry to advance technologies associated with human mobility and lunar surface infrastructure to support the Artemis missions. The Extravehicular Activity (EVA) Development Project within EHP manages the space suit development for the Artemis missions and is located at the Johnson Space Center.[5] • In March 2023, NASA established the Moon to Mars program office to oversee the programs contributing to the Artemis missions, including SLS, Orion, EGS, HLS, space suits, and the Gateway. This new program is responsible for end-to-end risk management and risk acceptance for the entire exploration system. The new program office resides within the Exploration Systems Development Mission Directorate. Prior to establishing the Moon to Mars program and during the course of our review, the Artemis Campaign Development (ACD) Division was responsible for overseeing the Artemis III mission integration. The Moon to Mars program is leveraging previous work performed by ACD on integrating across the programs to support the Artemis missions.
Contracting Approach for Human Landing System and Space Suit Development	To support the Artemis III mission, NASA awarded firm-fixed-price indefinite delivery, indefinite quantity contracts to two companies, SpaceX and Axiom Space.[6] SpaceX is to develop and demonstrate the human landing system, and Axiom Space is to develop the space suits. In July 2021, NASA exercised a $2.9 billion option on its contract with SpaceX to

---

[5] The EVA Development project also manages space suit development for the International Space Station, which is outside the scope of this report. For the purposes of this report, references to EHP and its officials encompass the EVA Development project.

[6] Under a firm-fixed-price-type contract, the government pays a fixed price regardless of actual cost and places upon the contractor maximum risk and full responsibility for all costs and resulting profit or loss. A firm-fixed-price contract provides for a price that is not subject to any adjustment on the basis of the contractor's cost experience in performing the contract. It provides maximum incentive for the contractor to control costs and perform effectively and imposes a minimum administration burden upon the contracting parties. FAR 16.202-1.

provide crew access to the lunar surface and demonstrate initial capabilities required for deep space missions.[7] Subsequently, in May 2022, NASA awarded firm-fixed-price indefinite delivery, indefinite quantity contracts to Axiom Space and Collins Aerospace. These companies are to provide safe and reliable commercial extra vehicular activity services in microgravity and partial gravity environments on the International Space Station and the lunar surface for Artemis missions.[8] In September 2022, NASA issued a $229 million order under Axiom's contract for the development and demonstration of a suit for lunar surface activities. Axiom is required to provide space suits that will allow crew to successfully perform exploration and science missions on the lunar surface during the Artemis III mission.

NASA is acquiring the human landing system and space suits as services, which represents a relatively new contracting approach for NASA.[9] NASA's intent is to transition from the traditional government-owned hardware model to a service model. Using this approach, the programs set high-level requirements and rely on the contractor's innovation to develop and deliver the system. Contractors only receive payment after NASA determines that the contractor has successfully achieved a milestone as defined in the contract. NASA is moving toward using this service model contracting approach for some acquisitions on its Artemis missions because the agency believes that it increases competition, innovation, flexibility, speed, and affordability.

Previously, NASA used a service model contract approach for its Commercial Crew Program (CCP). We reported in 2019 that NASA awarded firm-fixed-price contracts to SpaceX and Boeing in 2014 to

---

[7]NASA first awarded the HLS contract to three providers in May 2020. In April 2021, NASA announced the selection of SpaceX for the award of the contract to develop the Artemis III human landing system. After the award, Blue Origin and Dynetics filed bid protests with GAO, which GAO denied in July 2021. GAO, *Blue Origin Federation, LLC; Dynetics, Inc.-A Leidos Company,* B-419783; B-419783.2; B-419783.3; B-419783.4, July 30, 2021, 2021 ¶ CPD 265 (Washington, D.C.: July 30, 2021). Subsequently, in August 2021, Blue Origin filed a complaint with the U.S. Federal Court of Claims. The court dismissed this complaint in November 2021. Blue Origin Fed. LLC v. United States, Fed. Cl., No. 21-1695C (Nov. 4, 2021).

[8]An indefinite delivery, indefinite quantity contract provides for an indefinite quantity, within stated limits, of supplies or services during a fixed period. The government places orders for individual requirements. FAR 16.504(a).

[9]The acquisition strategy for the HLS program involves procurement of initial landing capabilities from each provider as Research and Development per FAR Part 35. Recurring services per FAR Part 37 will be procured in later acquisitions.

acquire human space flight transportation services to and from the ISS for CCP.[10] Under the contracts, we reported NASA also evaluated whether the contractors met its requirements and certified contractor final systems for use. In 2020, NASA determined that one of the CCP contractors, SpaceX, met the agency's standards for human space flight and certified it to conduct crewed missions to and from the ISS. As of May 2023, the other contractor, Boeing, was still working toward certification.

Prior to awarding the contracts for the human landing system and space suits, NASA took steps to reduce risk and increase industry participation in the programs. For example,

- To inform the human landing system development, NASA conducted risk reduction studies on topics including cryogenic fluid management and landing systems. Through these studies, NASA worked with several companies to get feedback on proposed human landing system requirements, develop element designs, and identify key technologies.

- NASA made its space suit design available to potential contractors and published a technical library with details of the design on the NASA website. NASA documentation states that the agency made this information available to reduce cost, schedule, and technical risks and contractor barriers to entry for providing a space suit for the Artemis III mission.

- The HLS program and EHP sought lessons learned from prior commercial human space flight programs, including CCP, to inform their acquisition strategies for the human landing system and space suits. Between 2019 and 2022, multiple NASA centers and mission directorates hosted knowledge-sharing events to facilitate the sharing of lessons learned from programs that used commercial service-type contracts. HLS and EHP officials attended these events.

Key Elements of the Artemis III Human Landing System

The human landing system will provide crew access to the lunar surface and demonstrate initial capabilities required for deep space missions. SpaceX is currently developing a commercial Starship vehicle to transport humans and cargo to low-Earth orbit, the moon, and Mars.

---

[10]GAO, *NASA Commercial Crew Program: Schedule Uncertainty Persists for Start of Operational Missions to the International Space Station*, GAO-19-504 (Washington, D.C.: June 20, 2019).

The HLS Starship system consists of the SpaceX Super Heavy booster (launch vehicle) and HLS Starship (the vehicle that provides crew access to the lunar surface). The HLS Starship is based on a common Starship architecture and shares many of the same critical systems, including propulsion, structures, and avionics, with the commercial Starship. The Raptor engines—which require liquid methane and liquid oxygen (collectively referred to as propellant)—power both the Super Heavy booster and the HLS Starship. In addition to the HLS Starship, SpaceX is developing a propellant tanker and on-orbit propellant depot—also variants of the commercial Starship vehicle—for its lunar landing mission concept.

SpaceX's plan for landing NASA astronauts on the moon for Artemis III includes multiple steps conducted sequentially. The order of these steps is as follows:

1. The propellant depot will be launched to low-Earth orbit, followed by multiple tankers that will rendezvous with, dock to, and transfer propellant to the depot;

2. Once sufficient propellant is on-orbit, an uncrewed HLS Starship will launch into low-Earth orbit, then rendezvous with and dock to the depot. The depot will transfer its propellant to the HLS Starship. The HLS Starship will then perform a rapid transfer into near-rectilinear halo orbit, where it will loiter for up to 90 days to confirm vehicle health and await the launch and arrival of Orion (the 90-day time frame is to accommodate any potential Orion or SLS launch delays);[11]

3. Orion will then launch with crew on board and dock with the HLS Starship;

4. Two astronauts will transfer from Orion into the HLS Starship, which will descend to the lunar surface for a 6.5-day stay; and,

5. Once the lunar surface activities, including moonwalks, are complete, the HLS Starship will ascend back to near-rectilinear halo orbit, where the crew will transfer back to Orion for their return to Earth.

Figure 2 depicts SpaceX's mission concept for the Artemis III lunar landing.

---

[11] Near-rectilinear halo orbit is a 1-week lunar orbit balanced between the Earth's and moon's gravity. This orbit enables global lunar access and promotes access to the lunar poles.

Figure 2: SpaceX Mission Concept for Human Landing System (HLS)

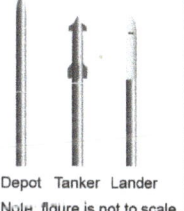

SpaceX's depot Starship launches from Earth and then loiters in low-Earth Orbit awaiting fuel. The tanker Starship will deliver fuel to the depot through multiple flights for transfer in space.

Once the depot accumulates sufficient fuel, the HLS lander launches and docks with the depot for fuel transfer. After the HLS tank is filled, it will decouple from the depot and transition to a near-rectilinear halo orbit around the moon awaiting the Orion capsule.

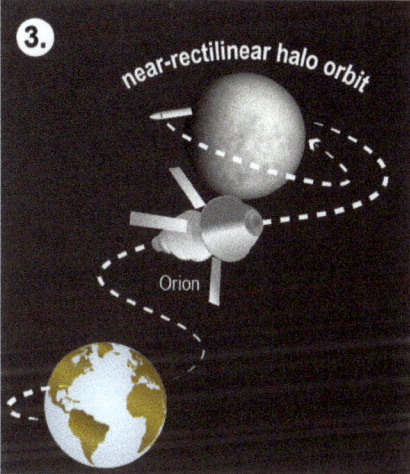

The Orion capsule will launch from Earth with mission crew onboard and rendezvous with the lander in near-rectilinear halo orbit around the moon.

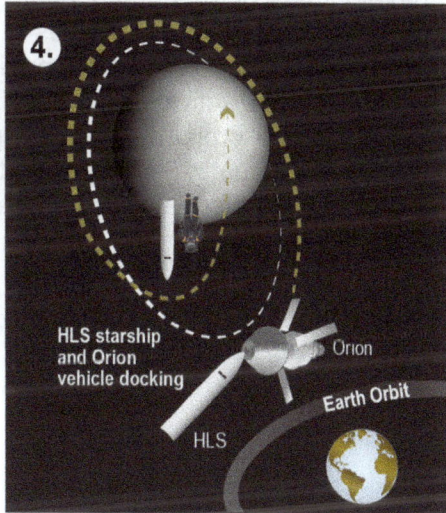

Orion will dock with the lander and after safety checks are complete, the crew will transfer from Orion to the lander. Then the vehicles will decouple and HLS will fly to and land on the moon's surface.

HLS crew will conduct science and exploration activities during a limited stay on the moon's surface. At the end of the lunar stay, the lander will leave the moon and rendezvous with Orion for crew transfer and final departure to Earth in the Orion capsule.

Source: GAO presentation of NASA and SpaceX information. | GAO-24-106256

Key Elements of the Artemis III Space Suits	Modernized space suits and associated hardware will provide portable life support, as well as tools for crew to use for lunar science and maintenance tasks. In December 2021, after spending 14 years and $420 million on next generation space suit development, NASA determined it would acquire modernized space suits from industry rather than develop the space suit system in-house.[12] NASA worked to mature several aspects of its design and, as noted above, made that design available publicly to help enable industry development of a space suit system. NASA's design is known as the government reference design. To deliver and demonstrate lunar surface space suits and associated systems, Axiom is leveraging many aspects of NASA's government reference design. The space suit consists of a combination of a pressure garment and life support components that together will provide capacity for at least 8 hours of lunar surface activity. The in-space system will consist of a variety of tools, flight support equipment, and other hardware that enable lunar surface exploration activities.
Life Cycle of a NASA Space Flight Project	NASA is using its life-cycle review process to oversee the contractors' development of the human landing system and the space suits. This process consists of two phases— (1) formulation, which takes a project from concept development to preliminary design, and (2) implementation, which includes activities like building, launching, and operating the system. Major projects must get approval from senior NASA officials at key decision points (KDP) before they can enter each new phase. Figure 3 depicts NASA's life cycle for space flight projects.

---

[12]NASA Office of Inspector General, *NASA's Development of Next-Generation Spacesuits*, IG-21-025 (Washington, D.C.: Aug. 10, 2021).

Figure 3: NASA's Life Cycle for Space Flight Projects

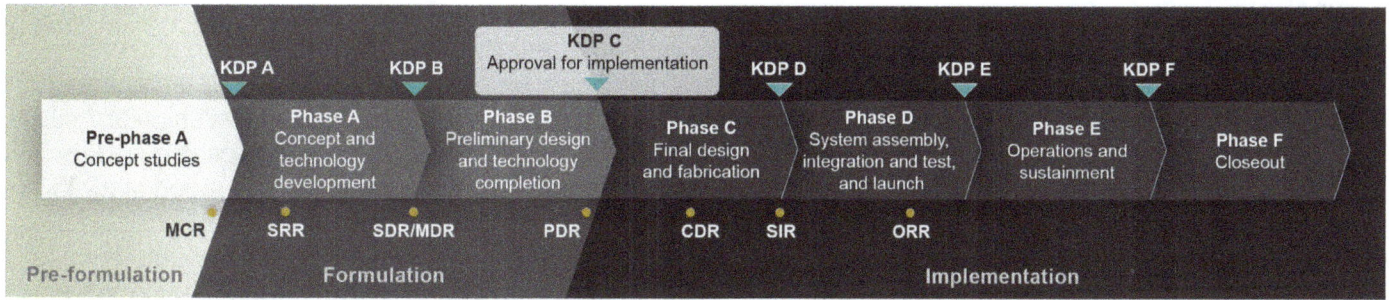

Source: GAO presentation of NASA images and information. | GAO-24-106256

In phase A, a project team develops a range of cost and schedule estimates for uses such as budget planning. During phase B, the project team develops programmatic measures and technical leading indicators that track various project metrics such as requirements changes, staffing demands, and mass and power utilization. Near the end of formulation, leading up to the preliminary design review (PDR), the project team completes technology development and its preliminary design.

Formulation culminates in a review at KDP C, where senior leaders approve the cost and schedule agency baseline commitments. After a project holds KDP C, it begins implementation, consisting of phase C where the project team holds a critical design review (CDR) to determine whether the design performs as expected and is stable enough to support proceeding with the final design and fabrication. As of July 2023, the HLS program is approaching KDP C while the EVA Development project—i.e., the space suit project—is approaching PDR.

2025 Crewed Lunar Landing Is Jeopardized by Delays and Scale of Remaining Work

NASA and SpaceX completed several important milestones for the human landing system since our September 2022 report, but a variety of factors make a lunar landing in 2025 unlikely. NASA officials are currently reviewing the HLS schedule. Axiom is making progress on the space suits, but it also has significant work to complete before the planned 2025 launch. At the same time, NASA and the HLS and Extravehicular Activity and Human Surface Mobility (EHP) programs are addressing a wide range of cross-program risks to landing humans on the moon in 2025.

Limited Lander Progress Makes 2025 Launch Unlikely Given Delays and Remaining Work

SpaceX has made progress in designing and testing components of its HLS Starship which, as discussed above, is the human landing system that will provide crew access to the lunar surface. However, the contractor is facing multiple issues that limit this progress and jeopardize its ability to support an Artemis III mission in 2025. These issues include an ambitious schedule, delays to key events, an incomplete orbital flight test, and a large volume of remaining technical work.

Ambitious Schedule

We found that if the HLS development takes as many months as NASA major projects do, on average, the Artemis III mission would likely occur in early 2027. Our analysis of past NASA projects that have launched since 2010 found that the average development time from project start to launch was 92 months. NASA has already delayed the Artemis III mission to December 2025, extending the HLS development time to 79 months. However, this is still 13 months faster than the average development time for NASA major projects. The complexity of human spaceflight suggests that it is unrealistic to expect the HLS program to complete development more than a year faster than the average for NASA major projects, the majority of which are not human spaceflight projects.

While SpaceX and NASA are aiming to complete development more than a year faster than the average for NASA major projects, they are achieving key events at a slower pace. For example, we found that SpaceX used more than 50 percent of its total schedule to reach PDR in November 2022. On average, NASA major projects used about 35 percent of the total schedule to reach this milestone.

Furthermore, the HLS program is taking longer to reach KDP C—the next key review after PDR—than average for the NASA major projects we assessed. As a result, the HLS program is proceeding with development without formal approval of a cost and schedule baseline. Specifically, the HLS program plans to use nearly 14 percent more of its total schedule to proceed from PDR to KDP C while on average NASA major projects used just 4.2 percent more of their schedule to achieve KDP C.[13]

Figure 4 illustrates the percent of total schedule used in achieving PDR and the percent that will be used in achieving KDP C by the HLS program, compared to the average for a NASA major project in our dataset.

---

[13]For more information on how we completed this analysis, see appendix I.

Figure 4: Human Landing System (HLS) Program Used a Greater Schedule Percentage to Achieve Planned Key Reviews than the Average for NASA Major Projects Launched since 2010

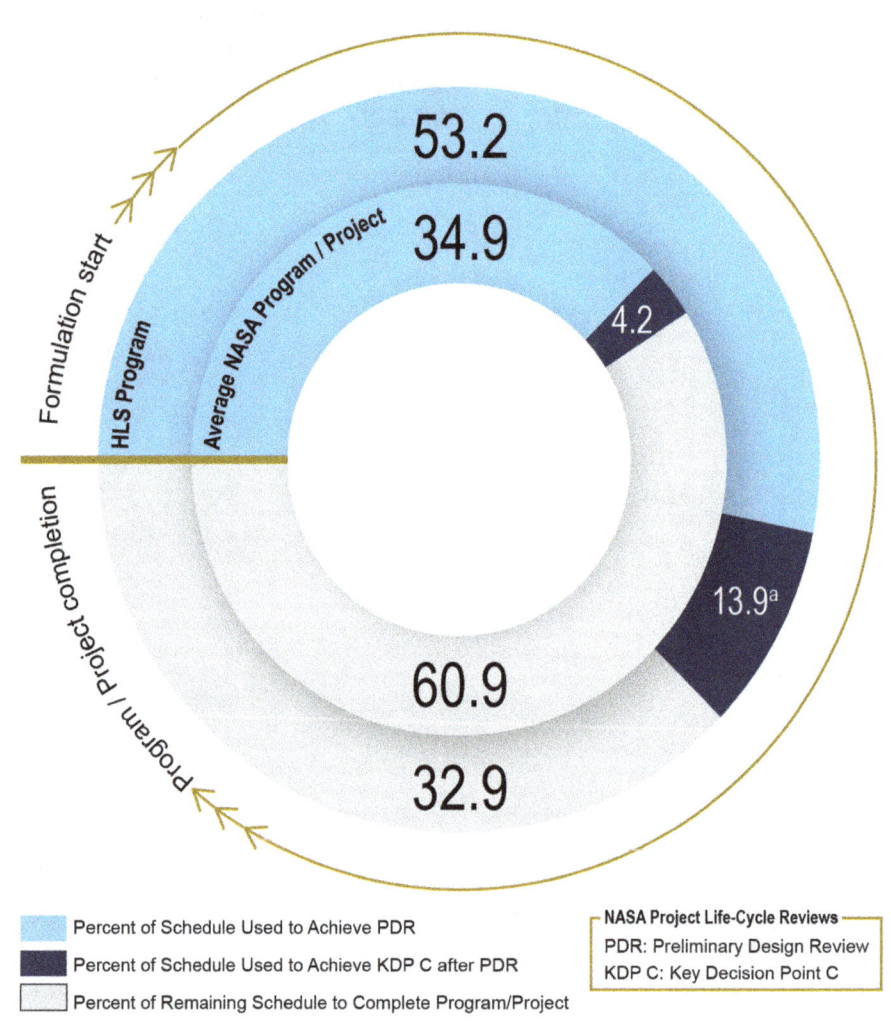

Source: GAO analysis of NASA data. | GAO-24-106256

[a]This calculation assumes the KDP C review will occur in October 2023, as currently planned, but that date was still in the future as of the writing of this report.

Delays to Key Events

SpaceX has also delayed several future program events that further compress the schedule. Since July 2022, the HLS program office and SpaceX delayed multiple key events from 2023 to 2024, meaning that many critical demonstrations and reviews will need to occur in the next 2 years to support an Artemis III mission as planned in 2025.

SpaceX and NASA continue to make progress on the human landing system, including completing some work early. SpaceX representatives reported completing 20 interim HLS milestones since June 2022 to mature the human landing system design and reduce development risk. NASA officials stated SpaceX submitted deliverables early for approximately 74 percent of the Artemis III contract payment milestones that have been completed.

Overall, the HLS program and SpaceX delayed eight out of 13 key events by between 6 and 13 months. Of those delayed events, at least two will occur in 2025—the year the Artemis III mission is scheduled to take place. Partially as a result of these delays, SpaceX plans to complete eight key events between November 2023 and the planned date of the Artemis III mission.

Due to delays to several key events, NASA will have a relatively short amount of time to ensure that the HLS complies with human spaceflight safety requirements before the mission start. For example, NASA delayed the HLS Design Certification Review, which is now closer to the Artemis III mission than originally planned. At this review, NASA will ensure that the design complies with requirements and human spaceflight certification. According to NASA documentation, this milestone should be completed 9 months prior to launch. As of September 2023, NASA and SpaceX also planned to complete Flight Readiness Reviews for the depot, tanker, and lander versions of the Starship within the same 9-month period before Artemis III. NASA and SpaceX will have less time to address any issues identified during these reviews before the mission. SpaceX representatives told us the three flight readiness reviews will always occur within the same length of time to support the Artemis III mission due to the nature and cadence of reviews. Any additional delays to these key events will compound the schedule pressure on NASA and SpaceX.

Incomplete Orbital Flight Test

In April 2023, after a 7-month delay, SpaceX achieved liftoff of the combined commercial Starship variant and Super Heavy booster during the Orbital Flight Test. But, according to SpaceX representatives, the flight test was not fully completed due to a fire inside the booster, which ultimately led to a loss of control of the vehicle. Following the launch, the Federal Aviation Administration—which issues commercial launch and re-entry licenses—classified the commercial Starship launch as a mishap and required SpaceX to conduct a mishap investigation. The Federal Aviation Administration reviewed the August 2023 mishap report submitted by SpaceX and, as a result, cited 63 corrective actions for SpaceX to implement before a second test.

SpaceX had planned this demonstration as the first test flight of the booster stage, as well as the first test with the Starship riding on the booster and the whole system experiencing stage separation.[14] However, SpaceX representatives said their Autonomous Flight Safety System initiated the vehicle self-destruct sequence and the vehicle began to break up about 4 minutes into the flight after the vehicle deviated from the expected trajectory, lost altitude, and began to tumble. HLS officials said that while the flight test was terminated early, it still provided data for several Starship technologies, including propellant loading, launch operations, avionics, and propulsion behavior.

The incomplete Orbital Flight Test led NASA to delay many key test events that are dependent on completion of that test. For example, NASA officials said that the in-space propellant transfer test will be delayed because it requires SpaceX to demonstrate that the Starship vehicle can reach orbit. Likewise, HLS officials told us that the Starship tests are sequentially linked, so future test flights, including the uncrewed flight test, depend on SpaceX successfully completing both the Orbital Flight and in-space propellant transfer tests.

A successful Orbital Flight Test is also needed to execute the technology maturation plan for the human landing system. For example, according to SpaceX representatives, the April 2023 Orbital Flight Test was key to demonstrating and understanding many aspects of the launch operations and ascent performance. SpaceX representatives said that they collected early in-flight data on the Raptor engines, vehicle tanks and primary structures, and pad and ground systems from the Orbital Flight Test. However, HLS officials stated that SpaceX is still required to perform a successful booster separation, ignite the Starship's engines, and achieve a suborbital altitude. HLS officials said that reaching orbit is essential for Starship development and that they expect that SpaceX's pace of design changes is likely to increase after a successful test flight, allowing them to make progress on finalizing the design and building hardware. SpaceX

---

[14]Stage separation occurs when the Super Heavy booster (first stage) and HLS Starship (second stage) disconnect after launch.

documentation states it plans to fly a second test in the fourth quarter of 2023.[15]

Lastly, NASA program officials said it is unknown whether the HLS Starship can be ready by the December 2025 launch date. The HLS program schedule required adjustments after the incomplete Orbital Flight Test and subsequent mishap investigation. In July 2023, NASA provided documents stating that the HLS schedule, including the program's ability to support a December 2025 launch date, are under review.

Remaining Technical Work

The HLS program will need to complete a significant amount of complex technical work on the engines and propellant transfer technology between 2023 and the end of 2025 to achieve the planned lunar landing goal. In a May 2023 NASA document, NASA officials overseeing the Artemis III mission integration stated that the HLS design maturity, with almost 3 years left before the planned launch, was insufficient. For example, they cited on-orbit propellant transfer as a potential issue because significant technical problems still need to be resolved.

The remaining technical work includes:

- **Raptor engine development.** SpaceX plans to use the Raptor engine in both the lander and booster stages of the human landing system and considers the technology to be relatively mature because it incorporates many years of prior development. However, the HLS Program Office identified engine development as a top risk to the program. SpaceX representatives said that their design for the Raptor engine follows an iterative approach, and as of September 2023, SpaceX had assembled and tested hundreds of engines. In a February 2023 interview, HLS officials said that if the Raptor engine operates below performance levels needed to meet mission requirements, thereby delaying engine certification, then it is possible that the new main engine for the human landing system will not be ready to support the planned mission in December 2025.

- **On-orbit propellant transfer technology.** SpaceX has remaining technical work to develop its on-orbit propellant storage and transfer technology. HLS program documentation states that propellant storage and transfer technologies have not previously flown in an integrated propulsion-like system. The documentation noted that, to

---

[15]SpaceX conducted the second Orbital Flight Test on November 18, 2023, and, this test was outside the scope of our review.

date, SpaceX has made limited progress in maturing those technologies.

There are multiple key systems related to the propellant transfer capability that SpaceX plans to develop for the human landing system. Those systems include docking sensors and mechanisms (to identify, locate, and then physically align the HLS Starship and the tanker Starship for fluid transfer); propellant measurement (to gauge the amount of propellant in the tanks and how much was transferred); and storage capability to mitigate fuel loss in space.

SpaceX plans to conduct the in-space Propellant Storage and Transfer test to further mature the technology, but the timing of this test is dependent on successful completion of preceding flights. SpaceX representatives said that the fundamental propellant transfer technology is not new or unique but requires engineering time and development effort to fully design and test for eventual use in the Artemis III mission. If the docking hardware does not perform as expected during spaceflight testing, significant vehicle modifications may be required, which could delay the mission. HLS officials said that SpaceX must demonstrate these technologies prior to completing the critical design review to promote confidence in the mission concept of operations.

Axiom Is Making Progress on Space Suits but Significant Work Remains

Axiom made progress in developing the space suits by completing several milestones, but it is still in the early phases of development. Since NASA's EHP gave Axiom authorization to proceed in September 2022, Axiom completed two NASA life-cycle milestones—Mission Concept Review in December 2022 and Certification Baseline Review in March 2023.[16]

In December 2022, Axiom built the first engineering development suit, derived from NASA's government reference design, with Axiom-constructed components. Axiom representatives said they use this development suit to test and redesign different components. In December, Axiom tested the suit at operational pressure in the lab using employees as in-house subjects to go through mobility exercise tests. Axiom also used the development suit to assess the sizing approach on test subjects. EHP officials said that Axiom is on schedule to complete

---

[16]The Certification Baseline Review is the milestone when the contractor establishes the design baseline, certification plan, life-cycle costs, and schedules for system certification. This milestone occurs early in the development phase and prior to PDR.

PDR in November 2023. Figure 5 shows Axiom's completed and remaining life-cycle milestones to deliver space suits to NASA for the planned December 2025 launch.

Figure 5: Completed and Remaining Milestones to Develop Space Suits for the Artemis III Mission

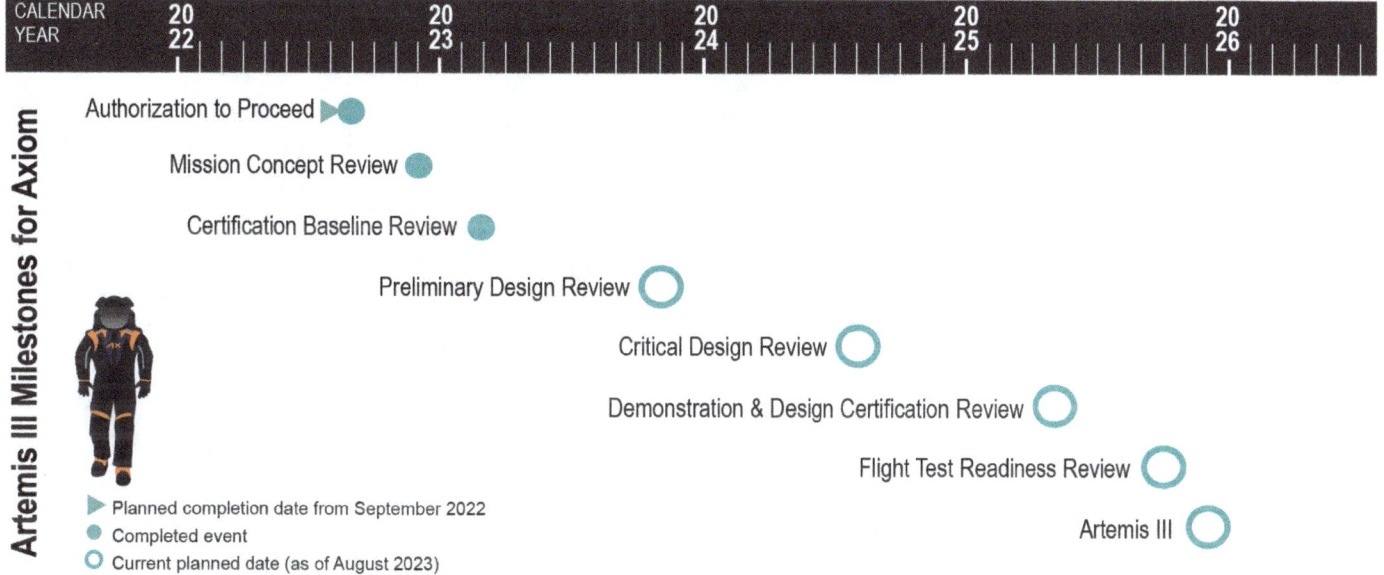

Source: GAO analysis of NASA and Axiom information. | GAO-24-106256

Axiom is also making progress on suit development by leveraging NASA's prior work. Axiom representatives said they brought in relevant experts and personnel who worked on the government reference design. These representatives said their approach was to adopt the government reference design and refine it to reduce costs to NASA.

At a February 2023 program review meeting with NASA, Axiom discussed its planned modifications to the government reference design space suit and its approach to mature several subsystem components in preparation for PDR in November 2023. For example, Axiom is repackaging the life support system to increase the size of the suits' oxygen tanks and building new components to improve NASA's government reference design. See figure 6 for an illustration of critical space suit subsystems and components that Axiom must mature by the planned December 2025 mission.

Figure 6: Illustration of Axiom's Space Suit and Major System Components

Source: GAO analysis of Axiom information and image. | GAO-24-106256

In addition, Axiom is modifying certain components of the government reference design to meet challenging requirements and address parts obsolescence issues.[17]

- **Emergency life support requirements.** NASA is requiring Axiom to develop a suit that can provide 60 minutes of emergency life support, more than any suit in history. Axiom representatives told us they may redesign applicable portions of the suit because the NASA government reference design did not satisfy the requirement to make the suit capable of storing that amount of oxygen. Axiom staff plan to decrease the size and rearrange components in the life support system package design to accommodate larger tanks that can hold more oxygen. However, if that is not possible, they plan to modify the current life support system, which Axiom representatives told us will take additional time.

- **Parts obsolescence and improvements.** Axiom representatives said they plan to incorporate, design, and certify new technologies—i.e., batteries, pumps, and electronic components—because designing new components helps them address supply chain and obsolescence issues. For example, Axiom is designing their space suit to be modular so that new technologies can be incorporated to address mission-specific requirements and allow incremental system upgrades. Axiom's remaining technical work on the space suits include modifications to all three sub-systems (Power, Avionics, and Controls; Pressure Garment System; and Life Support System).

Axiom is also making several improvements to the government reference design, which may require additional testing to mature. For example, Axiom considers one of the Axiom-designed life support components, the heat exchanger, as less mature than other suit components. However, Axiom representatives said their heat exchanger is outperforming the government reference design based on their testing. Additionally, NASA officials said Axiom is developing and building its own evaporative fibers for the space suit water membrane evaporator—a key system for cooling the suit's temperature. NASA officials said Axiom reported that modifications to the evaporative fibers are performing similarly, if not better, than the government reference design. NASA requested some of the new

---

[17]EHP officials said that the government reference design had limitations and challenges because it was not finished when NASA shifted strategies for providing space suits for the Artemis III mission. NASA officials said that the reference design suit had limitations because it was only matured to the CDR stage. At CDR, the design is mature enough to support full-scale fabrication, assembly, integration, and testing, but is still incomplete.

fibers Axiom proposed for the design to perform its own assessment but has not completed the assessment. As such, these component modifications could require further testing to mature.

In addition to resolving design issues for the space suits, Axiom has significant other remaining work to complete within the next 2 years:

- **Mature critical technologies.** Axiom plans to mature several critical technologies to meet mission requirements. For example, in January 2023, NASA assessed two critical space suit systems, the Life Support System and the Pressure Garment System, at a technology readiness level (TRL) 4.[18] Axiom's technology assessment rated over half of its critical technologies below TRL 6 and the lowest among those items at a TRL 3. One of those components, the Regenerable CO_2 Scrubber, is a life support system subcomponent that removes carbon dioxide from the suit environment. Axiom rated it at a low TRL since Axiom is not using the government reference design for this component.

    NASA officials told us they expect critical technologies to be at a TRL 6, which is mature, by the August 2024 CDR. In May 2023, EHP officials said neither the life support system nor the pressure garment system had significantly changed since January 2023—when they were both rated at a TRL 4—and those systems can only be matured through major testing at CDR. Prior to the November 2023 PDR, Axiom will complete the Crew Capability Assessment of the pressure garment system to further mature that technology. For CDR, Axiom will conduct human testing of the life support system in a NASA vacuum chamber facility.

    Axiom plans to produce and manufacture some components itself to reduce supply chain risks. However, NASA officials said that using a different source decreases the TRL for those components and, hence, the TRL will need to be reassessed. This is due to the uncertainty of a new manufacturer rather than a regression of the technology itself. NASA officials said the components that have design changes are also rated at a lower TRL and will need to be matured for use for the

---

[18]GAO, *Technology Readiness Assessment Guide: Best Practices for Evaluating the Readiness of Technology for Use in Acquisition Programs and Projects*, [Reissued with revisions on Feb. 11, 2020], GAO-20-48G (Washington, D.C.: Jan. 7, 2020). TRLs are a scale of nine levels used to measure a technology's progress, starting with paper studies of a basic concept (TRL 1) and ending with a technology that has proven itself in actual usage in the product's operational environment (TRL 9). A technology is considered mature when it reaches TRL 6, which is when a model or prototype is demonstrated in a relevant environment.

Artemis III mission. To support technology maturation efforts, Axiom personnel are developing multiple test rigs for different components of the life support system.

- **Procure suit components.** Axiom's remaining work to develop and procure suit components risks potential delays, which would compress Axiom's window to less than 2 years for delivering suits to NASA by September 2025. Prior to the Artemis III mission, the space suits need to be delivered so they can be integrated into the human landing system. Some of the parts that support critical systems for the space suit—including the life support system—have long lead times and could potentially take 12-18 months to procure from vendors. For example, Axiom plans to outsource certain parts, such as the oxygen regulators, because there are very few companies that make them to the standard needed for space suits. EHP officials said that procurement of components for the life support system is on Axiom's critical path for the schedule as some of these components are highly complex and specialized. EHP officials said that upon delivery, Axiom will have to conduct qualification and acceptance testing of those parts. To alleviate supply chain issues, Axiom now produces 40 of 61 parts in-house which gives it more control over the design and schedule for building the suit.

- **Qualify the suit for flight.** Axiom will qualify the space suits for flight readiness before a crew can use them, but the necessary testing facilities may not be available in time for the Artemis III mission. To certify the space suits for use, Axiom's proposed certification process requires the use of some NASA facilities. Axiom planned to conduct a crew capability assessment at the NASA Johnson Space Center's Active Response Gravity Offload System (ARGOS) facility as part of the PDR in November 2023. However, in August 2023, EHP officials said there was an issue with the equipment at the ARGOS facility. Axiom instead plans to test the space suits at the Partial Gravity Simulator facility at NASA's Johnson Space Center in October 2023, ahead of PDR. They also said that they are developing the supporting documentation so that the facility is prepared for Axiom to test the space suits. Additionally, Axiom will complete a vacuum test of the space suit at another NASA facility for CDR.

NASA Is Addressing Multiple Cross-program Risks That Affect Human Landing System and Space Suit Development

In addition to managing the remaining contractor work, NASA is also addressing multiple cross-program risks related to integrating the lander and space suits with systems needed for the Artemis III mission. Cross-program risks involve more than one program and require the programs to coordinate on the mitigation steps. These cross-program risks include, among other things, lander software and hardware integration and lunar dust contamination.

- **Lander software and hardware integration.** The HLS program is addressing a cross-program risk related to integrating the system's software and hardware with the Orion program. The HLS system uses software located across multiple hardware systems and subsystems, which makes it difficult to perform end-to-end software testing on flight-like hardware in relevant mission environments. HLS risk management officials said the program will need emulators and simulators to conduct HLS-Orion joint verification testing.[19] NASA officials said the Orion and HLS programs have made agreements to share four sets of emulator and simulator hardware and software and agreed to the details and exchange dates for each. However, in July 2023, NASA documentation stated that the HLS development pace does not align with Orion program integration milestones and could hinder the planned December 2025 launch readiness date.

  Integration of software developed for dissimilar hardware platforms, using different operating systems, as well as use and integration of heritage software, can be challenging and prone to introducing defects.[20] HLS risk documentation states that adequate test facilities and test campaigns are required to avoid late discovery of critical software defects because critical issues are often uncovered when software is integrated and tested with flight hardware. Therefore, without adequate testing, it is possible that critical software defects are missed. This situation could result in cost and schedule effects, or worse, potential loss of mission or crew.

  The HLS program and SpaceX have discussed applicable lessons learned from Boeing's Starliner first Uncrewed Flight Test conducted for CCP. This flight test failed to reach orbit due to software issues. The lessons learned included the need to increase the resources dedicated to software insight and oversight. The HLS program incorporated these lessons learned into guidelines on how NASA

---

[19]Emulators and simulators are software tools used to test critical systems under conditions and environments often unattainable in the lab or test bench. Simulators provide realistic data as would be experienced in actual flight while an emulator is a statistical approximation of the simulator.

[20]Legacy and heritage software are software products written specifically for one project and then, without prior planning during their initial development, found to be useful for other projects.

personnel should consider SpaceX-provided data on software development to ensure defects are identified earlier in development.

- **Lunar dust contamination.** Lunar dust is abrasive and is hazardous to the crew and equipment. For example, inhalation of lunar dust can cause irritation of the respiratory system and contact with the dust can cause irritation of the eyes. NASA's Artemis III program offices, including HLS, EHP, and Orion, are working to resolve a potential risk of lunar dust intruding into hardware and negatively affecting the performance of those systems. If not adequately mitigated early in the design process, it may cost more and take longer to retrofit the hardware to safely address lunar dust contamination.

Each program is working on its own portion of limiting dust exposure. For example, the HLS program is testing cleaning techniques for the lander's hatches and Axiom is developing tools to clean the space suit following a moonwalk. EHP officials expressed concern about the amount of dust that the space suits will pick up and bring into the HLS. Even if each system adequately meets its dust contamination requirement, there is still a risk that crew could become injured or ill due to the dust. Before the Artemis III mission, Axiom plans to demonstrate that it can limit the amount of dust on the exterior of the suit that is brought into the lander's cabin environment. SpaceX plans to demonstrate that the HLS Starship is capable of limiting lunar dust to a defined, acceptable level to safeguard crew health.

NASA's Moon to Mars Program Office—currently responsible for integrating all of the Artemis III systems—is also addressing performance and safety risks related to the known harm that lunar dust could inflict on the hardware and crew. It is addressing these risks across the HLS, space suits, and Orion programs based on lessons learned from the Apollo mission and subsequent research. Additionally, the HLS program coordinated with other NASA personnel and programs to develop a long-term mitigation plan focused on increasing their knowledge of south pole lunar dust and its properties.

NASA Is Taking Steps to Ensure Systems Are Safe and Meet Its Needs before Launch

NASA plans to take multiple steps to determine whether SpaceX's and Axiom's systems meet its mission needs and are safe for the crew. The HLS and EHP programs will ultimately determine whether the contractors' systems meet contract requirements. Then, NASA will conduct a to-be-decided series of reviews to determine whether the agency is ready for launch based on guidance it is currently developing. Further, NASA's contracts with SpaceX and Axiom grant NASA visibility into many areas of contractor work while the companies complete their significant remaining work.

NASA Will Determine If Contractors' Systems Meet Mission Needs, Including Safety

Before NASA conducts the Artemis III mission, it plans to determine whether SpaceX's and Axiom's systems meet requirements and are safe. NASA's requirements documents include requirements that the human landing system and space suits must meet. We have categorized these requirements into two types: system (includes functional and performance) and interface. System requirements include what functions the system needs to perform to accomplish the objectives and how well the system needs to perform the functions. Interface requirements describe how the HLS Starship interfaces with other systems—such as the Artemis space suits and Orion—so that each system will be able to safely operate with the other, among other things. See table 1 for examples of system and interface requirements.

Table 1: Examples of Requirements Applicable to SpaceX and Axiom, by GAO-Identified Type

	Human Landing System - SpaceX	Space suit - Axiom
Examples of system requirements	Be capable of operating on the lunar surface for a minimum of 6.5 Earth days In the event of an aborted attempt at a lunar landing, have the capability to conduct a safe return and dock to Orion in near-rectilinear halo orbit	Sustain the life of the crewmember for a minimum of 8 continuous hours of operation independent of vehicle-provided life support functions Provide safe and accurate visual capability and head mobility to perform tasks in both daytime and nighttime conditions
Examples of interface requirements	Provide oxygen to the space suit system at a temperature in the range of 40 to 90 degrees Fahrenheit for space suit system recharge Have a passageway providing for the transfer of crew and cargo to and from Orion	The space suits shall provide an interface for the receiving of oxygen for space suit recharge, in-suit pre-breath, and umbilical operations

Source: GAO analysis of SpaceX and Axiom contracts with the National Aeronautics and Space Administration (NASA). | GAO-24-106256

NASA's system requirements for the HLS Starship and space suits include NASA-approved alternative technical standards that the systems must meet:

- The HLS program allowed SpaceX to use alternative technical standards in three areas as long as NASA determined that the alternative standards met the intent of NASA's technical standards. These areas were safety and mission assurance, health and medical, and engineering technical standards. NASA officials said that after they completed the adjudication process with SpaceX, approximately 50 percent of SpaceX's technical standards were alternative to NASA's.

- Similarly, NASA allowed Axiom to propose alternative or tailored standards to NASA's design and construction, safety, and human health and medical standards in its initial proposal. EHP officials said they accepted nine tailored or alternative standards and determined that none of the changes created additional risk to NASA.

NASA officials from both programs reviewed the alternative technical standards early in system development. HLS officials said they completed this work early based on lessons learned from CCP to promote a shared understanding of technical standards and avoid late disagreements that could cause delays. EHP officials said that undergoing this process early in development was consistent with NASA policy and a good engineering practice to avoid taking unknown risks.

The programs and contractors each have a role in determining whether the HLS Starship and space suits meet NASA's requirements. SpaceX and Axiom will conduct activities—such as analysis, test, demonstration, or inspection—to verify that their systems meet NASA's requirements. The contractors will submit their verification results to NASA for approval or disapproval. For example, initial submissions are due at PDR for Axiom.

While NASA policy allows but does not require design certification reviews, SpaceX and Axiom are contractually required to undergo this review before the Artemis III mission to ensure that their designs comply with system and interface requirements.[21] At these reviews, NASA will review each contractor's design and evidence to ensure that the contractor's system meets all system and interface requirements, among other things.

NASA Is Formulating Plans to Determine Agency Readiness to Launch

In December 2022, NASA's Artemis Campaign Development Division released an implementation plan for Artemis missions that provides a high-level summary of NASA's planned approach to technical and programmatic reviews, among other things.[22] The implementation plan states that after each of the programs' design certification reviews, there will be a design certification review of the integrated architecture requirements for Artemis III. This review will assess whether the five-program integrated system—HLS, EHP, Orion, SLS, and EGS—can meet

---

[21]NASA Procedural Requirements (NPR) 7123.1C, *NASA Systems Engineering Processes and Requirements* (Feb. 14, 2020).

[22]The Artemis Campaign Development Division Implementation Plan was released before NASA announced that it established the Moon to Mars program office. A January 2023 NASA report on Moon to Mars program implementation stated that the program will leverage the work done by the Artemis Campaign Development Division and will update the existing implementation plan to reflect any necessary changes. This plan will generally apply to the Artemis III mission and beyond and the applicable programs.

requirements across all applicable configurations and environments, among other things. Then NASA will undergo a to-be-determined certification of flight readiness process to determine if these programs are ready to execute the mission (see fig. 7).

Figure 7: NASA Processes to Determine Whether a Contractor's System Meets Requirements before Key Reviews to Support Artemis III Launch

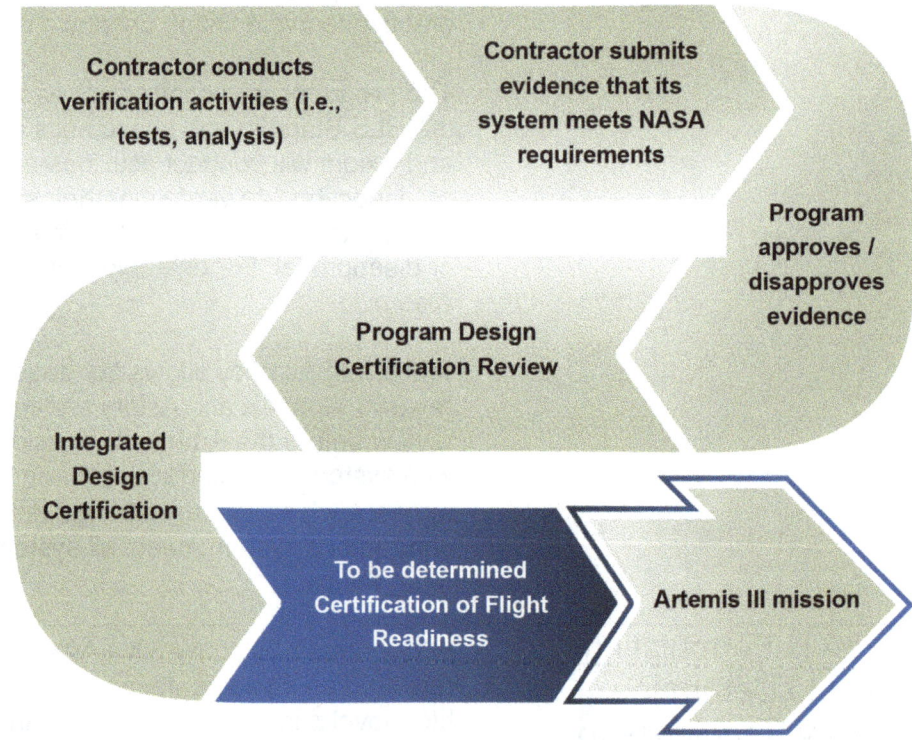

Source: GAO analysis of NASA, SpaceX, and Axiom plans and contract documents, and interviews with NASA officials. | GAO-24-106256

NASA officials said they are currently developing general guidance for implementing certification of flight readiness for future Artemis missions as well as Artemis III mission-specific guidance.

- NASA officials said the general Artemis certification of flight readiness guidance will likely establish a framework for programs to develop their own certification of flight readiness plans and define the reporting structure for Artemis flight readiness, among other things. NASA plans to obtain a formal, integrated human rating certification for the Artemis III mission and associated crewed space system architecture as part of the certification of flight readiness process. NASA policy states that a human-rated system is required for crewed space systems to control hazards, to manage safety risks associated with human spaceflight, and safely recover crew.[23]

- Officials also said the Artemis III mission-specific certification of flight readiness guidance will likely explain the timeline and sequence of multiple flight readiness reviews that will take place to determine readiness to launch.

The flight readiness review (1) examines tests, demonstrations, analyses, and audits that determine if a system is ready for a safe and successful flight or launch; and (2) ensures all flight and ground hardware, software, personnel, and procedures are operationally ready. Officials said they anticipate that the agency will hold at least eight flight readiness reviews to determine readiness for Artemis III:

- five program-level reviews (HLS, EHP, Orion, SLS, EGS),
- two Moon to Mars program-level reviews, and
- one agency-level review led by the Exploration Systems Development Mission Directorate Associate Administrator, who is responsible for assessing the flight readiness of programs and projects within the mission directorate.

NASA officials said they are applying NASA guidance and best practices from earlier human spaceflight efforts to their Artemis certification of flight readiness plans. These earlier efforts include the Space Shuttle, CCP, and ISS missions. For example, officials said that like ISS missions, the Artemis III certification of flight readiness process will begin around 3 months before launch. However, they said there are limitations with this approach due to the complexity of integrating five systems for the Artemis III mission. There are multiple launches that will take place for the Artemis III mission, which will include the Starship depot and tankers, as well as the HLS Starship and the crew on Orion. Officials said NASA will have to determine the timing and spacing of the reviews for the Artemis III mission, given the number of launches needed to execute the mission.

NASA officials said they plan to release both the general Artemis certification of flight readiness plan and the Artemis III mission supplement in 2024. They stated that they plan to release both plans 1

---

[23]NASA Procedural Requirement (NPR) 8705.2C, *Human Rating Requirements for Space Systems* (July 10, 2017).

year before launch to give them enough time to be implemented for Artemis III.

NASA's Contracts Grant Visibility into Technical Progress and Safety during Development

NASA included clauses in the contracts with SpaceX and Axiom to gain visibility into contractor efforts under the firm-fixed-price contracts for services. The SpaceX and Axiom contracts grant NASA visibility into contractor efforts throughout development and at key milestones as follows.

Approval authority. Both contracts require the contractors to submit specific data deliverables at key milestones and identify which deliverables must be approved by NASA. For example,

- Both contracts require a system safety assessment report that documents potential safety hazards and methods to control those hazards. The contractors must submit these safety reports for NASA review and approval at multiple milestone reviews to support program and independent safety review panels at NASA.

- Each month, both contractors are also required to submit an integrated master schedule that documents planned work, and resources necessary to accomplish that work. Axiom's monthly integrated master schedule submission must include a schedule risk analysis.[24] SpaceX must submit a schedule risk assessment and an integrated master schedule, at the critical design review and design certification review. SpaceX's integrated master schedule does not need to be approved by NASA, but it will be used to plan, manage, and report work required in performance of the contract. In contrast, NASA has a time-limited right to disapprove Axiom's integrated master schedule.

The HLS program manager said that while the program has yet to experience delays due to needing time to review SpaceX data, program staff recognize that timely review will be critical as SpaceX's development progresses. The program manager said that the program developed a schedule for reviewing SpaceX data to provide its technical opinion on a timely basis.

---

[24]Our schedule assessment guide defines schedule risk analysis as incorporating program schedule risks into a statistical simulation to predict the level of confidence in meeting a program's completion date; to determine the contingency, or reserve of time, needed for a level of confidence; and to identify high-priority risks. This analysis should be performed before a baseline is set. GAO, *GAO Schedule Assessment Guide: Best Practices for Project Schedules*, GAO-16-89G (Washington, D.C.: December 2015).

Insight. The SpaceX and Axiom contracts grant NASA insight or access into many areas of contractor efforts, including any changes that could affect the mission or crew safety. SpaceX and Axiom provide NASA staff insight through the use of recurring meetings, electronic systems, and access to their facilities. Both contracts also grant NASA insight into certain aspects of SpaceX's and Axiom's commercial variants of the Starship and space suits, respectively. For example, HLS officials said that insight allowed them to observe SpaceX's activities leading up to the Orbital Flight Test, which flew a commercial Starship variant. Both contracts include a government insight clause that lays out the scope of NASA's insight. See table 2.

Table 2: Scope of NASA's Insight into SpaceX and Axiom Activities

SpaceX contract	Axiom contract
Areas of insight include: • Any aspect of the design, development, analysis, testing, schedules, performance metrics, risks, and management processes of the contractor's human landing system and individual vehicles, elements, integrated systems, subsystems, etc. According to NASA, this includes every aspect of the integrated lander, all supporting spacecraft, and if applicable, any Active-Active docking adapter. • Launch vehicles and launch site operations. According to NASA, this includes, but is not limited to, spacecraft-to-launch vehicle integration, spacecraft handling procedures, launch commit criteria, and range safety analysis supporting the launch of the integrated lander elements, supporting spacecraft, and all payloads, including non-NASA payloads or cargo. • Flight and mission operations. According to NASA, this includes preparations, flight plans, rules and procedures, trajectory and mission design, crew and flight control team training, real-time operations, space communications and navigation networks, and any non-NASA services performed by the contractor. • Any other contract performance activities or data identified by NASA that are mission-critical or otherwise related to safety in any manner.	NASA has insight into Axiom, subcontractor, and partner entities' efforts that could affect Artemis requirements, interfaces, integration, operations, and crew safety. These include: • Design • Development • Manufacturing • Management • Mission integration • Vehicle integration • Operations • Medical and health • Training and certification • Hardware/software testing NASA has access to all Axiom activities associated with Artemis Interface compatibility and safety certification. NASA has access into any Axiom-initiated space suit changes across its commercial endeavors or any changes that may affect NASA missions.

Source: GAO analysis of SpaceX and Axiom contracts with the National Aeronautics and Space Administration (NASA). | GAO-24-106256

NASA officials said that the purpose of ensuring insight into contractor efforts is to verify technical information, which should help ensure that formal milestone reviews and deliverable submissions are successful. For example:

- HLS program officials said they were able to use insight clauses and data deliverables to gain visibility into how SpaceX's critical technologies were maturing, even though SpaceX is not required to provide TRL information. Further, they said that information exchanged through insight opportunities helps SpaceX focus its own design and documentation efforts, which should result in higher-quality deliverables for formal reviews.
- EHP officials said that they included insight clauses in Axiom's contract to ensure that NASA has extensive insight into the contractor's design and interfaces, including changes to NASA's government reference design. EHP officials said that insight into contractor activities will be used between formal milestones to maintain awareness of emerging issues on a timely basis. They believe this awareness will reduce the likelihood that Axiom's milestone reviews are delayed or put NASA in a position to rush the review.

As NASA and its contractors are learning to implement the insight clauses, NASA officials said they were cognizant that insight activities could pose a schedule risk to its contractors. For example:

- The HLS program manager said that conducting insight on a schedule-driven, firm-fixed-price contract was a culture shift for NASA staff who were used to conducting oversight on cost-reimbursement contracts.[25] By default, the HLS program is using its access granted by the insight clauses to primarily monitor SpaceX's efforts, but it also established processes to change the depth of its insight. For example, the HLS program could conduct independent analyses to corroborate its understanding of SpaceX's progress. While the program allows for insight levels to be changed based on an ongoing assessment of risks, any change in the level of insight must be approved by the HLS program manager. The program established this process in recognition that deeper insight levels would require additional resources from (1) the HLS program to conduct deeper insight, and (2) SpaceX to address NASA's insight results. HLS officials said that areas in which the program has deepened its insight to date include

---

[25]Cost-reimbursement types of contracts provide for payment of allowable incurred costs, to the extent prescribed in the contract. These contracts establish an estimate of total cost for the purpose of obligating funds and establishing a ceiling that the contractor may not exceed (except at its own risk) without the approval of the contracting officer. FAR 16.301.1. Cost-reimbursement contracts require the government to maintain extensive visibility into a contractor's technical progress and financial performance. FAR 16.301-3(a)(3)(4).

system safety, propulsion, software, flight mechanics, landing stability, and contingencies.

According to both HLS officials and SpaceX representatives, SpaceX gave NASA insight beyond what the contract requires by sharing information about its work under other programs as well as its Starship development. The program reported that its insight into SpaceX's early Starship development has been beneficial for NASA and SpaceX as they solve problems and reduce risk. SpaceX representatives explained that they were sharing information with NASA because these early efforts will inform what is needed for the HLS Starship. They also said they carried over "crew office hours" that were used with CCP, where NASA astronauts can have detailed discussions with the SpaceX team on topics such as mission operations.

- EHP officials said that Axiom's contract gives NASA the ability to dig deep into any area that poses risk and there is no barrier for NASA to seek access. Therefore, officials stated that insight activities represent schedule risk to Axiom because the contractor does not know how much time or how many staff will be needed to address or accommodate NASA's insight activities. EHP officials also said that they have already learned that insight processes were bottlenecked by administrative processes, such as how quickly they were able to share information, rather than by technical work.

It is too soon to determine how well the HLS and EHP programs will balance their insight into the contractors' efforts with the schedule pressure to conduct the Artemis III mission as planned in 2025. We previously reported that CCP's insight into contractor efforts took more time than the program or contractors anticipated.[26] More recently, an Aerospace Safety Advisory Panel report stated that NASA continues to have a strong safety culture.[27] Further, in May 2023, the panel chair told us that she does not have any concerns about NASA losing focus on safety, even as pressure to move faster to meet the Artemis III date increases.

Collaboration. NASA included clauses related to the use of government resources in both contracts to formally define the availability of NASA expertise to its contractors. However, the contracts differ in how

---

[26]GAO-17-137.

[27]The Aerospace Safety Advisory Panel was established by Congress to provide advice and make recommendations to the NASA Administrator on safety matters. NASA, *Aerospace Safety Advisory Panel Annual Report 2022* (Washington, D.C.: January 2023).

collaboration relates to approval authority and insight. The Axiom contract states that collaboration is the highest form of insight, which EHP officials explained to mean that knowledge gained through collaboration can be used to understand the system and its risks. In contrast, the HLS contract states that collaboration and insight are distinct and governed by uniquely applicable terms and conditions. Both contracts state that NASA has the sole authority to determine the amount, type, or duration of support it provides, and the support remains under the supervisory control of NASA.[28] HLS and EHP officials said the clauses were mutually beneficial to NASA and the contractors because they allow the contractors to leverage NASA's expertise while NASA experts get an opportunity to do hands-on work.

Since contract award, both companies are leveraging or planning to leverage NASA support. SpaceX can request up to 60 full-time NASA employees or equivalent support contractor personnel, while Axiom can request up to 25 full-time NASA employees during contract performance. HLS officials said it has provided 36 equivalent personnel to SpaceX to support specific tasks, such as preparation for the Orbital Flight Test. They said SpaceX also requested collaboration in areas such as micrometeoroid orbital debris, Raptor engine development, characterization of lunar landing sites, and risk assessment of the SpaceX-Starship Pad 39A at Kennedy Space Center. Axiom requested collaboration support in 25 areas such as manufacturing, lighting, and crew training.

Agency Comments

We provided a copy of this report to NASA for review and comment. NASA provided technical comments, which we incorporated as appropriate.

---

[28]SpaceX's contract also allows for NASA or the NASA support contractor to have supervisory control of NASA staff during collaboration, as appropriate. Axiom's contract states that NASA staff remains employed by, and under supervisory control of, NASA staff at all times during collaboration.

We are sending copies of this report to the appropriate congressional committees and the Administrator of NASA. In addition, this report is available at no charge on the GAO website at http://www.gao.gov.

If you or your staff have any questions about this report, please contact me at (202) 512-4841 or RussellW@gao.gov. Contact points for our Offices of Congressional Relations and Public Affairs may be found on the last page of this report. GAO staff who made major contributions to this report are listed in appendix II.

William Russell
Director, Contracting and National Security Acquisitions

Appendix I: Objectives, Scope, and Methodology

The objectives of our review were to describe (1) the extent to which the National Aeronautics and Space Administration (NASA) has made progress in developing key systems needed to land humans on the moon in 2025, and (2) the steps NASA is taking to ensure that its lunar landing systems contractors are developing systems that meet NASA mission needs and are safe for crew. This is the latest in a series of GAO reports addressing NASA's Artemis enterprise.[1] This report focuses on the human landing system (HLS) initial capability, as well as the Artemis Exploration Extravehicular Activity systems—referred to as space suits.[2]

To determine the extent to which NASA made progress in developing key systems needed to land humans on the moon in 2025, we reviewed lunar landing system programs' plans, contract documentation, and quarterly program status reviews to identify program milestones and critical technology demonstrations. The HLS program, Extravehicular Activities (EVA) and Human Surface Mobility program (EHP), and EVA Development project are overseeing the contractor-led development of these systems. For both the HLS and the space suits, we selected milestones from contract and non-contract sources to track program progress for developing these systems. To select milestones, we identified test flight events and technology demonstrations from contract documents, NASA policies on program life-cycle reviews and project management, and schedule information collected from the HLS and EHP programs. We compared actual and planned schedule milestones for HLS as of June 2022, the earliest date that data were available, to July 2023 data. We compared actual and planned schedule milestones for the EVA Development Project—i.e., the space suit project—as of the contract award in September 2022 and compared them to the latest information available as of September 2023.

---

[1]GAO, *NASA Lunar Programs: Improved Mission Guidance Needed as Artemis Complexity Grows*, GAO-22-105323 (Washington, D.C.: Sept. 8, 2022); *NASA Lunar Programs: Significant Work Remains, Underscoring Challenges to Achieving Moon Landing in 2024*, GAO-21-330 (Washington, D.C.: May 26, 2021); *NASA Human Space Exploration: Significant Investments in Future Capabilities Require Strengthened Management Oversight*, GAO-21-105 (Washington, D.C.: Dec. 15, 2020); and *NASA Lunar Programs: Opportunities Exist to Strengthen Analyses and Plans for Moon Landing*, GAO-20-68 (Washington, D.C.: Dec. 19, 2019).

[2]In this report, we focused on landing systems needed for Artemis III, specifically the human landing system initial capability and the Artemis space suits. NASA is also pursuing a sustained lunar landing capability and modernized space suits for the ISS from industry. However, we did not include these efforts because they are not included in the Artemis III mission.

Appendix I: Objectives, Scope, and Methodology

To examine the planned development time frames for the HLS program relative to time frames for other major NASA projects, we calculated the number of months from program start to completion for these programs and compared the average number to HLS planned time frames. We used cost and schedule data collected for our prior work assessing selected major NASA projects and reviewed data reliability assessments completed for that work. Based on our review of data reliability assessments for that work, we determined the data are reliable for our purposes as the relevant data, time frames, and analyses were the same. We also calculated major projects' percentage of schedule use for key reviews, which included preliminary design review (PDR) and key decision point (KDP) C for projects that were included in our assessment of major project reports and launched between 2010 and 2022.[3] In addition to the HLS program, we analyzed 29 spaceflight projects in our dataset, and excluded non-spaceflight projects completed within the above time frame.

We determined the scope of remaining development work for both programs to meet the planned Artemis III mission date in December 2025 by reviewing program and contractor schedule milestones, risk charts, plans for mitigating risks, and technology maturation plans. Additionally, we analyzed these documents to determine the current operational and technical risks facing lunar landing system acquisitions and assess the remaining work for the programs and contractors. We interviewed officials from the HLS and EHP programs, as well as SpaceX and Axiom personnel, to understand the status of the lunar landing systems programs.

To determine what steps NASA has taken to ensure the contractors will deliver systems that meet mission needs and ensure crew safety, we assessed HLS and EHP program documentation and SpaceX and Axiom contract and system requirements documentation. We collected and reviewed contracts and associated attachments that outline statements of work and requirements and contractor verification and validation plans, as well as NASA policy on human spaceflight safety. We interviewed HLS and EHP program officials, which included the EVA Development project and the Commercial Crew and Cargo program officials. We also

---

[3]GAO, *NASA: Assessments of Major Projects*. GAO-23-106021 (Washington, D.C.: May 31, 2023); and *NASA: Assessments of Major Projects*, GAO-22-105212 (Washington, D.C.: June 23, 2022). When NASA determines that a project has an estimated life-cycle cost of over $250 million, we include that project in our annual review up through launch or the project's end of development.

Appendix I: Objectives, Scope, and Methodology

interviewed SpaceX and Axiom personnel. We reviewed SpaceX and Axiom insight implementation plans and interviewed NASA and contractor personnel regarding contract execution. We also reviewed NASA agendas and presentation slides on lessons learned from using a service contract approach, as well as mission integration documentation. We interviewed NASA Artemis Campaign Division officials to understand their certification of flight readiness plans for the Artemis III mission.

We conducted this performance audit from September 2022 to November 2023 in accordance with generally accepted government auditing standards. Those standards require that we plan and perform the audit to obtain sufficient, appropriate evidence to provide a reasonable basis for our findings and conclusions based on our audit objectives. We believe that the evidence obtained provides a reasonable basis for our findings and conclusions based on our audit objectives.

Appendix II: GAO Contact and Staff Acknowledgments

GAO Contact	William Russell, (202) 512-4841 or RussellW@gao.gov
Staff Acknowledgments	In addition to the contact named above, Kristin Van Wychen (Assistant Director); Erin Roosa (Analyst-in-Charge); John Armstrong; Susan Ditto; Lorraine Ettaro; Nathan Hanks; Tonya Humiston; Erin Kennedy; Joy Kim; Meredith Allen Kimmett; Douglas Luo; Edward J. SanFilippo; Sylvia Schatz; Kate Sharkey; Kevin Walsh; and Alyssa Weir made significant contributions to this report.

Related GAO Products

NASA: Assessments of Major Projects. GAO-23-106021. Washington, D.C.: May 31, 2023.

NASA Lunar Programs: Improved Mission Guidance Needed as Artemis Complexity Grows. GAO-22-105323. Washington, D.C.: September 8, 2022.

NASA Lunar Programs: Moon Landing Plans Are Advancing but Challenges Remain. GAO-22-105533. Washington, D.C.: March 1, 2022.

NASA: Lessons from Ongoing Major Projects Could Improve Future Outcomes. GAO-22-105709. Washington, D.C.: February 9, 2022.

NASA Lunar Programs: Significant Work Remains, Underscoring Challenges to Achieving Moon Landing in 2024. GAO-21-330. Washington, D.C.: May 26, 2021.

James Webb Space Telescope: Project Nearing Completion, but Work to Resolve Challenges Continues. GAO-21-406. Washington, D.C.: May 13, 2021.

NASA Human Space Exploration: Significant Investments in Future Capabilities Require Strengthened Management Oversight. GAO-21-105. Washington, D.C.: December 15, 2020.

NASA Commercial Crew Program: Significant Work Remains to Begin Operational Missions to the Space Station. GAO-20-121. Washington, D.C.: January 29, 2020.

NASA Lunar Programs: Opportunities Exist to Strengthen Analyses and Plans for Moon Landing. GAO-20-68. Washington, D.C.: December 19, 2019.

NASA Commercial Crew Program: Schedule Uncertainty Persists for Start of Operational Missions to the International Space Station. GAO-19-504. Washington, D.C.: June 20, 2019.

GAO's Mission	The Government Accountability Office, the audit, evaluation, and investigative arm of Congress, exists to support Congress in meeting its constitutional responsibilities and to help improve the performance and accountability of the federal government for the American people. GAO examines the use of public funds; evaluates federal programs and policies; and provides analyses, recommendations, and other assistance to help Congress make informed oversight, policy, and funding decisions. GAO's commitment to good government is reflected in its core values of accountability, integrity, and reliability.
Obtaining Copies of GAO Reports and Testimony	The fastest and easiest way to obtain copies of GAO documents at no cost is through our website. Each weekday afternoon, GAO posts on its website newly released reports, testimony, and correspondence. You can also subscribe to GAO's email updates to receive notification of newly posted products.
Order by Phone	The price of each GAO publication reflects GAO's actual cost of production and distribution and depends on the number of pages in the publication and whether the publication is printed in color or black and white. Pricing and ordering information is posted on GAO's website, https://www.gao.gov/ordering.htm. Place orders by calling (202) 512-6000, toll free (866) 801-7077, or TDD (202) 512-2537. Orders may be paid for using American Express, Discover Card, MasterCard, Visa, check, or money order. Call for additional information.
Connect with GAO	Connect with GAO on Facebook, Flickr, Twitter, and YouTube. Subscribe to our RSS Feeds or Email Updates. Listen to our Podcasts. Visit GAO on the web at https://www.gao.gov.
To Report Fraud, Waste, and Abuse in Federal Programs	Contact FraudNet: Website: https://www.gao.gov/about/what-gao-does/fraudnet Automated answering system: (800) 424-5454 or (202) 512-7700
Congressional Relations	A. Nicole Clowers, Managing Director, ClowersA@gao.gov, (202) 512-4400, U.S. Government Accountability Office, 441 G Street NW, Room 7125, Washington, DC 20548
Public Affairs	Chuck Young, Managing Director, youngc1@gao.gov, (202) 512-4800 U.S. Government Accountability Office, 441 G Street NW, Room 7149 Washington, DC 20548
Strategic Planning and External Liaison	Stephen J. Sanford, Managing Director, spel@gao.gov, (202) 512-4707 U.S. Government Accountability Office, 441 G Street NW, Room 7814, Washington, DC 20548

Please Print on Recycled Paper.